Horst Berger

Light Structures
Structures of Light

The Art and Engineering of Tensile Architecture
Illustrated by the Work of Horst Berger

1663 Liberty Drive, Suite 200
Bloomington, Indiana 47403
(800) 839-8640
www.AuthorHouse.com

© 2005 Horst Berger. All Rights Reserved.

No part of this book may be reproduced, stored in a retrieval system, or transmitted by any means without the written permission of the author.

First published by AuthorHouse 06/07/05

ISBN: 1-4208-5267-1 (sc)

Library of Congress Control Number: 2005903812

Printed in the United States of America
Bloomington, Indiana

This book is printed on acid-free paper.

To
my family and my students

Table of Contents

Preface to the Second Edition IX

Preface and Acknowledgment X

Foreword XII
Two outstanding professionals introduce the subject: Tony Robbin is an artist and mathematician with a special interest in structures. His book, entitled "Engineering a New Architecture", is an excellent introduction to the topic. Mamoru Kawaguchi is one of the most prominent contemporary structural engineers. His work includes the roof structure for the Olympic Arena in Barcelona.

1. Introduction - Building a Gentler Environment 1
This chapter sets the tone by introducing the subject, stating the objectives, and summarizing the essence of what is to come. Throughout history, architecture has been achieved by a marriage of technology and art. Indeed, technology and the language of contemporary art are intimately linked to the past and the future. Though each individual project featured was designed in response to a specific set of architectural requirements, as a group they represent an attempt to explore forms which the technology of the new age might offer, in order to create a built environment that is efficient, gentle, respectful of nature, and delightful to live with.

2. In the Beginning: Domes 9
People built structures as shelter for almost half a million years. In so doing, they developed the structural forms which have dominated architectural design throughout its history. A few examples demonstrate the origins of these very early forms, and trace them through history.

3. From Tents to Tensile Architecture 21
Tensile structures, in the form of tents made of natural materials, go back more than thirty thousand years. People on the move, such as nomadic tribes or armies, have used them throughout recorded history. The industrial age, with its new high-strength materials, first used them to built long-span bridges in the early 1800s, but the earliest true tensile structure covering an architectural space is only 50 years old.

4. How Tensile Structures Work. . . . 41
Tensile Architecture requires structure to have priority in creating the form of the building. Compared with external loads (snow and wind), the weight of the structure and enclosure is infinitesimal. The components of the structure are flexible, so that gravity and rigidity cannot be marshalled to generate stability and load carrying capacity in the structure; form and the condition of stress are relied on instead. This is a new concept in architecture.
  This chapter reviews the elementary principles which underlie the form generation of tensile structures, and describes the benefits of their large deformation behavior, using simple diagrams.

5. Materials for Tensile Structures. . .55
An exploration of fabric properties shows the impact of the membrane on the function and appearance of the building. Strength, durability, translucency, reflectivity, heat absorbtion, and acoustical properties are discussed. The properties of other structural materials are also reviewed.
  New materials have entered since the book was first published, hinting at a significant future for fabric structures in architecture

6. 20 Years of Fabric Structures: 1973 to 1993 69
This chapter is based on the author's direct experience with the design, engineering, construction, and performance of fabric tensile structures, and includes the following sections:

6.1 The Way of the Spider 75
The early development of radial tent structures.

6.2 Celebration in the City 85
Bicentennial structures in Philadelphia, and the roof for Canada Place in Vancouver.

6.3 A Forest in the Desert 95
The design and construction of the Haj Terminal at the Jeddah International Airport, the world's largest roof structure.

6.4   Big Tops to Stay 113
Pole supported structures are the simplest tensile forms. This section describes a series of buildings which use this primary structural form to enclose space.

6.5   Moving the Masts Outside. 121
A-frames are an efficient and elegant way of avoiding the penetration of interior space by pole supports. This section describes a series of tent-like structures suspended from various arrangements of A-frames.

**6.6   A Giant Flower
          Shades the Stadium** 133
The roof cover over the Riyadh Stadium is a third larger than any other stadium roof, in spite of its large central opening. The huge petals of its flower-like folds give it the strength and stability to withstand desert storms. Its structure is simple and easy to understand. Its construction process was amazingly elegant.

6.7   Stretched over Slender Arches . . 143
Using arches of wood, concrete or steel, elegant and useful spaces were created. The basic configurations are in principle very similar to the first human shelters.

6.8   Music under the Tent. 153
The Mitchell Performing Arts Center near Houston has become a favorite place for audiences of up to 10,000 to listen to a symphony concert or a musical in the area. The structure provides intimacy for the auditorium beneath, and forms a friendly proscenium arch for the lawn audience outside.

6.9   Sails for San Diego 163
The fabric roof of the San Diego Convention Center covers the 100,000 sq. ft. outdoor exhibit area. Its simple structure spans the space without interior supports, and allows the ends to be free and open. The sail-like structure has become a landmark for this city.

6.10   Bringing Light to the Airport . . . 171
The design and construction process of this most important public use of fabric architecture is described as part of the most advanced and impressive airport design. The airport was one year old when this book first appeared and has its ten year anniversary when this edition comes out..

7.   Covering Very Large Spaces.. . . . 187
Fabric structures are superior in their capacity to achieve very large spans. This chapter describes executed long span roofs, and concepts for rigid and movable roof systems.

**8.   Physical and Mathematical
        Models** 199
The critical step in the development of tensile structure designs is the generation of form. A lack of familiarity with the methods of handling form generation is the greatest obstacle to the use of fabric structures. This chapter introduces the two most important design tools: the use of stretch fabric models and the use of mathematical modeling. As is shown, they are not difficult to handle, especially with the use of a new form-finding method.

9.   Ideas and Possibilities 215
A number of concepts and designs are briefly discussed.

10.   Project Credits 225
Credits for participation in the design and construction of the main featured projects are given in tabular form.

11.   Illustration Credits 227

Preface to the Second Edition

When I first began to write this book I was a consulting engineer deeply enmeshed in the competitive world of building design. For the last 15 years I have been mainly a teacher in a school of architecture, and I see the world from a different angle.

Nine years have passed since this book was first published. Although it received praise from professionals, teachers and lay persons, the original publisher did not think a second printing profitable. So I decided to take a different tack. In order to make the book affordable to a wide audience, I undertook the task of reformatting the book, so it could be published by a print-on-demand publisher at a very low price. It is one of the first of its kind with its more than 400 illustrations. This time-consuming labor is now behind me, and the book is here in this simple, black and white format. And I am pleased with the result.

I am aware that the book may not be as carefully edited as it could be, and I ask for your indulgence. Write me, if you find errors to be corrected.

With this much time passed, it became necessary to edit and revise the book. But I kept this to a minimum. So there might be a passage here or there which may sound a bit dated as a result. However, it was more important to retain the integrity and originality of the story.

It is apparent to me now, that the structures I describe in the book are more unique than I had thought. Although fabric structures have been built since, none of them seems to have the architectural significance of this series of structures. And I reluctantly admit that I might have something to do with this. I am also convinced that fabric tensile structures will re-emerge in the architectural future. There are indicators, at least to my optimistic mind.

Good architecture will not remain limited to luxury projects. Translucent fabric enclosures provide affordable answers to the design of delightful and useful architectural space. Fabric enclosures are more economical than glass enclosed structures; they require less maintenance; and they are more desirable for environmental reasons.

There are also signs that the obstacles to fabric structure design and construction are beginning to be overcome. User friendly design programs - including one shown in Chapter 8 of this book - will help in the design process, especially if they fit into the new digital process which every student of architecture now grows up with. And new fabric materials are beginning to appear - see Chapter 5 - which might drastically reduce the cost and simplify the handling of this critical element.

I believe that architecture - and the built environment in general - will have a brilliant future in this post-industrial century. I say this in spite of the irrational wars we are in right now. Because I am an older man now, I have a longer memory. The world has come a long way since the devastation of World War II. If we come to our senses, we can make it the beautiful space we have been dreaming of for so long. But it will always take commitment and dedication.

Hastings-on-Hudson, March 2005

Preface and Acknowledgements

This book tells the story of the design and construction of a series of fabric tensile structures which have occupied my life for several decades. They were part of the evolution of tensile structures into a new architectural technology and a new building form. I wrote the book from my perspective as a participant, and I tried to capture some of the excitement and the anxieties of the process. Because fabric architecture is a new phenomenon, I took pains to explain the principles which determine its form and the mechanisms which make it work.

I explored its roots in ancient forms of human dwelling and traced its use throughout history.

I introduced basic design procedures and I offered a glance at potential future applications in various parts of the built environment. In all of this I took great care to make the book accessible and interesting to readers who have no technical training without compromising the quality of information to students and professionals in architecture and engineering.

My opinions and biases are, no doubt, quite apparent. I hope they give life to the story rather than detract the reader. I am not a scientist. I was educated as a civil engineer with the clear objective of using science to make this a better environment for people to live in. My mind is that of an integrator who tries to bring things together, my instincts are those of an artist who sees the world as a unit. I have had the great fortune of spending my professional life in the wonderfully positive and creative world of construction in which ideas are converted into realities, and to find a field of activity in which my presumably divergent talents could become useful.

That my title is civil engineer rather than architect is an obvious source of consternation to many and I must, therefore, insert some comments on this subject. Though civil engineers became a distinct scientifically-trained profession, apart from architects, early in the 19th century, their role in the design and construction process is still little known to the general public. Their primary activity is devoted to public work structures such as railroads, highways, airports, canals, and water supply systems. As structural engineers they became the primary consultants to architects, designing and analyzing the structural frames which make buildings stand up. In fact, the new structural forms they invented, the materials they helped develop, the scientific methods they generated were critical contributions which made contemporary architecture possible.

As a young man I was equally fascinated by art and technology. Though I graduated as a civil engineer, the decision whether to be an architect or an engineer remained unmade. The tensile structures described in this book are - at least in part - the product of this unmade decision. It is my hope and my belief that in the not too far future this segregation between art and technology will disappear and the true architect - the artist who masters technology as a means of creating space for human activity - will be recreated. This is one reason why I am devoting this part of my life to teaching students of architecture.

Just as the buildings which this book describes were created by a multitude of people, I had the generous and able support by a number of people who assisted me in the writing of this book and to whom I want to extend my thanks: Mamoru Kawaguchi, the brilliant structural engineer who has designed some of the most daring and elegant structures in recent years, has encouraged me throughout the process and has contributed a foreword; the other foreword comes from Tony Robbin, artist, mathematician, writer, and connoisseur of structural forms who has just published his own book on structures in architecture. Tony's enthusiasm has been enormously helpful to me.

Norval White, distinguished architect, teacher, and writer, has acted as advisor and has reviewed the book with his fine sense of style. My friend William Wood applied his unerring sense of language with immense energy and thoroughness in editing the first draft of the book. With his background as a business executive he is responsible for making the book understandable and readable for those not trained in the building professions. Above all, Bill taught me to write English. Ilse Seidel, my sister, applied her critical sensitivity of language to the final version of the text.

Light Structures Design Consultants, especially my partner Joseph DeNardis together with my teaching assistant Mimi Kueh, produced the CAD drawings which are an important part of the book's illustration.

Birdair Inc. of Buffalo, New York, the fabricator and contractor for most of the fabric structures presented in this book, provided most valuable support in many ways. Without Birdair's assistance the rich photographic illustration would not have been possible.

There are many others who have been of assistance. Ulrike Ruh of Birkhäuser has been most patient and helpful in the complex process of making a book. John Dennis Gartrell applied his skills to give the text its final touch, and Gregor Messmer turned the layout into a piece of art.

Most of all I am indebted to my wife, Gay, who helped in many capacities. She was the final editor of the book and tried to keep me orderly. With immense patience she took second place to "the book" over a long period of time.

I consider it a great fortune to have had the opportunity of participating in the evolution of the structures which are the subject of this book. I am joyfully aware that the light which they bring into the world does not originate with me but comes from a deeper source.

Hastings-on-Hudson, February 1996

Foreword by Tony Robbin

"One of the the first... Certainly the biggest... Surely the most beautiful..." These words describe some of the forty tensioned membrane structures designed by Horst Berger. Their great beauty is due not simply to a careful consideration of formal issues and fine detail, although Berger is sensitive to both. Rather, both the beauty and the practicality of these structures stems from a new concept of space that is implicit in them, and directs choice and informs taste - a point of view that saves these marvelous new buildings from being mere technological innovations or vacuous formalisms.

Horst Berger's favorite is the graceful Cynthia Woods Mitchell Pavilion for the Performing Arts, the summer home of the Houston Symphony completed in 1990. The membrane is supported by a three member A-frame truss that passes through it, and is thus both inside and outside the pavilion. The A-frame forces a double-curved surface that maximizes and main-tains tension in the membrane, giving it stability. Seating is both under the membrane and on a grassy berm; we are folded into the building at the same time as being returned to the outside. Such a sensation is reinforced by the purpose of the building, which is to project sound: the subtle, complex curves allow the musicians to hear themselves, and also allow the music to be focused on the audience without hot spots. Light also takes this path, first around the interior, and then out to the audience. The form, the structure, the function, and the mechanical workings of the structure are therefore one.

This inside out, Mobius strip of a building is a paradigm of the new sense of architectural space, made possible by membranes, and made real for us by Horst Berger. The inside is not claustrophobically walled off from the outside, as in some medieval fortress. Rather, membranes communicate between inside and outside; they are membranes in the biological sense: passing through or filtering out different elements from both sides. Membranes can be engineered to pass or reflect light, heat, sound, or moisture in whatever combination of directions one chooses. Horst Berger is among the pioneers who have realized that this new material requires a new shape, and that the new shape requires new construction methodologies and engineering techniques; but far more importantly, he has also seen that the new material requires the abandonment of antiquated concepts of architectural space. Architectural art is not at war with architectural science; the converse is true: new art is only delivered to us by new engineering.

An imperative tone is part of Horst Berger's philosophy: we have no right to waste solar energy, to make old fashioned solid walls to trap air that giant machines, huffing and belching like locomotives in the Gare St. Lazarre, alternatively heat and cool. Such illogical use of materials and resources offends his moral sense. In his lectUres, he often notes that the population of the earth has more than doubled in his lifetime, and as a father he feels protective of the generations that will follow. Membrane structures are Horst Berger's way of leaving a legacy of wealth to society, and more generally, of instructing us that rational choices can solve our problems, if only we are willing to let go of obsolete archtypes.

Horst Berger is in his mid sixties - smooth skin, bright eyes, silver hair, radiating energy, a quick step, a ready smile, and an informal manner. His own house, where he and his wife raised their four children, is a rambling adventure of additions to an old house, without a right angle in sight. Horst Berger is a committed teacher, a bench scientist, a hands on model maker. Although among the most respected engineers in the world - the one man to call when it becomes apparent that long-span space frame designs are impossible to build on time and on budget - Horst Berger is free of rigidity and conceit, and determined to stay light-weight, not bogged down in academia or professional administrative hassles. His personality is characterized by a rich mixture of seriousness and youthful fun. Because he has avoided the pretensions that destroy creativity, the work shown here is just the beginning.

Tony Robbin, New York, March 1995

Foreword by Mamoru Kawaguchi

The history of structural engineering can be regarded as a history of man's pursuit of ever lighter structures. Massive masonry arches and domes, weighing several tons per square meter, were for a long time the only means of spanning large building spaces. In the 19th century, the advent of iron and steel enabled engineers to reduce the weight of such roofs to a few hundred kilograms per square meter. A similar lightness in concrete, using thin shell construction, only became possible in the 1920s, however. Further development in structural engineering accelerated the trend towards yet lighter structures, such as space frames, cable structures, and finally, fabric membranes. Advanced computers, sophisticated software, and their recent rise in popularity have made possible the introduction of structures into the built environment which are literally "light", with weights below ten kilograms per square meter.

Light structures are special structures. They are special not only because of their lightness, but also because within them, little, if anything, separates configuration from formal expression. Their structural components - poles, cables, struts, roof, wall surfaces and all connecting joints - are visible both from outside and inside, in a composition that is necessitated by their function: namely, the transmission of forces. This is in direct contrast to "ordinary" structures, where structural components are usually hidden behind finishing elements, and whose textures and forms are considered more comfortable to the human eye and tactile senses. Although light structures may be painted, galvanized, or plated, the structure itself, being clearly visible, becomes the architecture of both form and space.

This inevitably leads us to the question: "Can a structure be beautiful by itself?" In art and architecture, this question has been asked repeatedly and answered varyingly by artists and architects alike, according to the changing tastes of the times. During the predominance of MODERNISM, the answer was a resounding "Yes"; after the public's captivation with POST-MODERNISM, it became a disinterested "No". Such fashionable fluctuations were promoted by those with a rather superficial attitude, and after unsuccessful attempts at producing new and interesting work by means of a POST-MODERN orientation, architects now seem to be re-addressing the question, discussing it with renewed interest; among serious people, however, the answer has always been "Yes".

Some readers of this book may recall that in his *Razon y ser des tipos structurales*, Eduardo Torroja, a talented Spanish structural engineering pioneer, wrote that structure can be a very effective means of architectural expression. But Torroja went further than mere statement; he also expressed this idea through his built designs, such as the Zarzuela racetrack grandstands and the aqueduct at Alloz. Those works showed that pure structure can be beautiful without added decorative features. It is interesting to note that Torroja built both structures in the 1930s, long before such a statement was of social interest. Pier Luigi Nervi, one of Torroja's contemporaries, eloquently demonstrated an excellence of structural expression in many built works, and showed how the economy and rationality of the construction process could contribute towards building a "correct" structure.

In the 1950s and 60s, when MODERNISM was predominant, many talented architects and engineers, among them Felix Candela, Eero Saarinen, Kenzo Tange, Frei Otto, Paul Weidlinger, Anton Tedesko, Yoshikatsu Tsuboi, and Heinz Isler, followed the pioneers Torroja and Nervi. And it should be noted that although, in the POST-MODERNIST era, architects became less interested in an expression of structural rationality, engineers continued to pursue the creative expression of rational structural forms.

Membrane structures, introduced into building engineering by Walter Bird as air-supported domes in the late 1940s, were continuously and ingeniously developed from the 1960s through to the 80s. Through his invention of low-profile air-supported domes, his introduction of Teflon-coated fiberglass fabrics, and his cable dome designs, the late David Geiger was among those who made important contributions to the development of membrane structures. Horst Berger, the author of this book and the former partner of Geiger, played an equally important role in the development of membrane tension structures. While Geiger mainly concentrated on air-supported structures, Berger worked on fabric tensile structures, and their introduction into a wide range of permanent architecture.

Berger is a man of genius, with a wide range of talents. As I suggested above, the answer to the question "Can a Structure be Beautiful by Itself?" should always be "Yes". That does not necessarily mean that a structure will automatically be beautiful. A beautiful structure can only be created by a designer who combines great sensitivity and imaginative richness with the ability to understand the behavior of that structure technically. Horst Berger is indeed a rare example of such a designer. Whenever I talk with him or listen to his lectures on tensile structures, I admire his enthusiasm, the purity of his ideas, and the clarity of his thinking. This is born out in fine examples of his work, such as the Ryadh Stadium roof; the structures of the San Diego Convention Center and Canada Place in Vancouver; the Mitchell Performing Arts Center near Houston; and the world's largest enclosure structure covering the terminal of the new Denver Airport.

When writing this foreword, I was fortunate in having an opportunity to read a few chapters of the book in galley form, and found them very, very attractive. I very much look forward to reading the rest of the book when it is published.

Mamoru Kawaguchi, Tokyo, January, 1995

Creating, yet not possessing;
Working, yet not taking credit.
Work is done, but not dwelt on.
Because it is not dwelt on
   nobody can ever take it away.

Lao Tsu, Tao Te Ching

1

Introduction - Towards a Gentler Built Environment

The roof of the Jeppesen Terminal at the Denver International Airport. August 1994 [1.01]

As I sit down to write this introduction it is just a few days since my wife and I, together with a group of friends, returned from a trip to the Rocky Mountains and my mind is filled with powerful images of nature: jagged mountain ranges; turquoise lakes which mirror them; glaciers, waterfalls and rocky streams, pathways of the soft water which shapes hard mountains and fills receptive lakes; and upright thrusting trees, whose magnificent structures were there long before us and will be there long after we are gone.

On the way out we stopped in Denver. John Gagnon, a young architect, drove us out to the new Denver International Airport. As a member of C.W.Fentress J.H.Bradburn and Associates, the Denver architectural firm in charge of the final design of the terminal building, John had played an important role during the design and construction of the project and we had made this trip together many times before.

Usually buildings open before they are completed. The new Denver International Airport was an exception; construction was completed in Spring of 1994. But, though several months had passed, the facility was not yet open due to problems with its electronic baggage handling system.

Being in a new building for the first time after its

Inside the Denver Terminal Building - August '94. From right: John Gagnon, my wife Gay, and I. [1.02]

completion is always a strange experience for me. Here the sensation was reinforced by the fact that the space was almost deserted. Nine months before, at the time of my last visit as design and engineering consultant, this had been a construction site, bubbling with activity. The process of converting ideas into reality had still been alive. Problems had to be solved, questions answered, drawings interpreted, details inspected, corrections requested, installations approved. Concepts had long before turned into sketches, models, drawings, specifications, contracts. Most building components had been fabricated and assembled. But the process had still been in flux and with it remained the sense of identification, even ownership, at least of the fabric roof and glass wall structures whose primary forms and elements had at one time existed in my mind only.

Now the building was finished and had taken on a life of its own. Soon it would belong to the travellers who would rush through the light filled hall on their way from one point in their busy lives to another, taking in - at least for a fleeting moment - the joy of this space, the discipline of its cathedral-like organization, the splendor of its huge glass walls, the gentle sweep of its weightless roof. And they might sense the deep connection this structure has to the forms of our nomadic past and the daring thrust its novel order heralds towards the new world of the coming millennium with its uncertainties and its bright potential.

Vancouver, which our group of friends reached towards the end of our vacation, confirmed and broadened the Denver experience. Though the exposure to nature on the grand scale of the Rocky Mountains had made human structures look insignificant, Vancouver proved to be a delightful city to visit. And Canada Place, with its sail-like fabric structure reaching out into the harbor, was clearly a prominent feature of its built environment. I had not seen the building since its completion nine years before. Due to the particular circumstances at the time I had not even been there during its construction. Therefore my memory was still engaged with the design process.

I vividly remember the heady one-day design session in architect Eb Zeidler's office in Toronto in which the concept of this roof structure was born. I recall rushing back to New York to make a stretch fabric model in order to test and demonstrate the feasibility of the structure I had dared to propose, and to have a design tool for us to work with the following week. On my way back to Toronto I carried the model in my tennis bag, worried that the Canadian custom officials would not let me enter - as they had threatened the week before - since I did not yet have a work permit for Canada. But nobody noticed.

That was all more than a decade ago. Now I was looking at the real building, long the property of its city and of the people that used it. It was larger than I had imagined it, its form stronger, richer, deeper. And it had nothing to do with me anymore.

The purpose of this book is to describe these and a series of other fabric tensile structures Which have occupied my life over the last two and a half decades. Some forty of them have been built; many more were designed. Most of those built are part of permanent buildings.

Introduction: A Gentler Built Environment

Canada Place, Vancouver, B.C.
August 1994 [1.03]

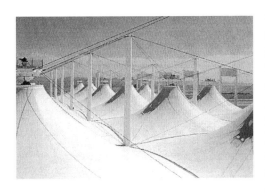

One of the ten modules of
the Hai Terminal roof at the
Jeddah airport during construction - Feb. 1980.
For scale note the worker in
front of the third mast in the
left background. The total
roof covers 105 acres of
land. [1.04]

Some have even become important landmarks of their communities.

Some are very large. The Haj Terminal roof at the airport in Jeddah, Saudi Arabia, is the largest roof structure in the world. The roof over the Riyadh Stadium is big enough to fit the Houston Astrodome inside the circle of its support columns. Yet some of my favorite designs are tiny.

That I was instrumental in their existence - as designer, engineer, promoter, model maker, dreamer, in roles ranging from head of the design team to one of its many specialty consultants - is only one aspect they have in common. There are more objective features they all share and which define them.

Dominant among them is the characteristic form which sets these structures apart and is a necessary aspect of their capability to carry load. Being made of light and flexible materials such as woven, coated fabric, and high strength cables, they lack gravity and rigidity which are the basis of structural strength in conventional building support systems. Instead, curvature and internal tension - similar to that which determines the pitch of a drum or holds the shape of a soap bubble - are the properties which make a tensile structure capable of resisting loads.

To keep a curved surface in tension while it carries upward loads from wind and downward loads from snow requires a careful balance of curvatures in opposite directions. The curved surface needs to be stretched between support points above the surface, below it, and on all sides around its periphery. These geometric requirements are common to all tensile surface structures and give them the shape by which they are instantly recognized. Curvature, following scientific laws, is a critical aspect of the structure. Consequently, the shape of the structure cannot be arbitrary but derives from the structural function. Form and function are one. Art and engineering are inseparable.

The continuous fabric surface which is an integral part of the structure, serves also as the enclosure, separating the interior space from the outdoor environment, keeping out water and wind, reflecting most of the sun's heat, transmitting natural light, controlling the intrusion and reverberation of sound.

One integrated structure, therefore, does all the things which in conventional buildings requires the combination of many additive elements, demanding the expert attention of numerous specialists for their design. And this clear and purposeful shape of the structure defines the space it encloses and gives the building its identity. The structure becomes its architecture.

This integrated approach to building design is largely foreign to the segregation and specialization which pervades the design and construction process today, a result of the technological and scientific advances of the industrial age. It resembles much more the the ways of the pre-industrial architect.

The builders of the great cathedrals of the Middle Ages knew no separation of art and technology, form and function, structure and construction. They were *masons*, from the master down to the least of the stone cutters, thoroughly trained in all aspects of transforming stone into magnificent buildings which would stand forever.

The building forms they used had their distant origin in the house structures of pre-historic communities, going back hundreds of thousands of years, examples of which we still find in the indigenous architecture of Africa and Asia and in the dwellings of the American Indians. They were built and owned most commonly by the women in the community, using easily available natural materials and employing sturdy structural systems, buildable in one day. In recognition of the deep formative impact they have had on our minds on what we expect building structures to look like, this book starts with a brief review of these early domes and their resonance through the history of architectural structures. It then explores the evolution of tensile structure forms, which began with nomadic tents.

These intuitive structural systems include many of the forms we use in contemporary tensile surface structures. But when expanding these primary structural forms to large and super-large spans enclosing major permanent building facilities, the methods of the pre-historic craftsman are no longer sufficient. In fact, even the tools and methods of the industrial age are not adequate for handling the design of advanced tensile surface structures. Their form follows complex geodesic space patterns rather than classical geometric shapes and they undergo large deformations under wind or snow loads.

These behavior patterns can no longer be simulated by the mathematical tools derived from classical physics. They require us to enter the new physics called *chaos*, which deals with non-linear phenomena, such as turbu-

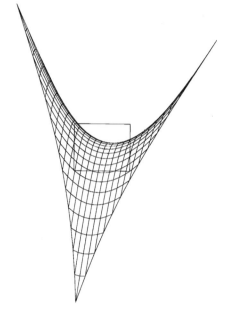

Common to all tensile structures is their anti-clastic geodesic surface: two sets of opposing force lines intersecting at right angles. [1.05]

The great cathedrals of the Middle Ages show no separation of art and technology, form and function, structure and construction. [1.06]

Introduction: A Gentler Built Environment

Sketch of a proposed convention center roof structure. Two-dimensional images are rarely sufficient for the design of tensile structures.[1.07]

Many of my structures began with making a stretch fabric model. Using the hands draws on deeper layers of the mind. [1.08]

lence, weather patterns, irregular heart beats, and fractals. And they cannot be analyzed without the new power tool, the electronic computer.

Drawings made on a flat drafting board are not sufficient to explore the potential of tensile structure forms and develop designs for buildings which use them. The book, therefore, describes principal aspects of two design approaches.

Stretch fabric models which behave very similar to the actual structure are by far the most useful way of learning about these structures, especially for beginners Most of my designs started that way. I believe that involving the hands in the design process draws on deeper parts of the brain than the more abstract processes we commonly use.

A section on mathematical methods of finding the equilibrium shape of tensile structures concentrates on an iterative computation method. The method is simple enough that the shape of small structural grids can be computed with the help of a small calculator. Most recently I have developed a user friendly, design oriented formfinding program which allow architects and engineers, or anyone interested, to develop surface structure forms.

The main part of the book describes actual structures, what makes them work, and how they came to exist. It is limited to projects I was personally involved in, simply because I know them intimately and they express best the values which are important to me. Obviously, these projects were not conceived as part of a body of work; to me, each of them was a job with its own specific circumstances and problems. But the same guiding principles were always behind the design of each structure: that it produce useful, meaningful, and delightful spaces for people; that it be affordable and buildable; that it minimize the exploitation of natural resources and do the least harm to the environment; and that its very nature become visible as a thing of beauty, as a manifestation of a deeper order of things, a more permanent light which, for just a short while, we are entrusted to keep shining in this world.

These principles seem to echo what the Roman architect Vitruvius, almost two thousand years ago, considered the primary qualities of a building: utilitas (function), firmitas (structure), venustas (beauty). They are still proclaimed by most practitioners of architecture. But in this eclectic period of transition between the Middle Ages and a new era of which we begin to see the outlines, the pursuit of these qualities is often lost in the confusion caused by an over-complex design and construction process with its enormous time and money pressures and with its numerous participants who share few common values. While the magic of the new technologies of the industrial age seems to have liberated architecture from the narrow confines of pre-industrial construction technology, it has taken away the clarity of form which graced historic and pre-historic architecture.

I believe that this clarity has to be re-discovered and regained to return strength of form to architecture. Tensile surface structures are not "the answer". But they demonstrate some of the ingredients which I believe are critical. The construction and operation of our buildings and our civil engineering structures today greatly contribute to the harm we inflict on the planet which sustains our lives. In order to survive with our ever increasing numbers, we must learn to change the violent ways in which we build and live.

The brute-force-energy-waste approach of the industrial age, with its relentless devastation of natural resources, must give way to gentle means of transforming the sun's abundant gift of energy and to the obedient attention to the laws of the natural environment. Aggressive competition must make space for joyous cooperation. Sharing must take the place of dominating. Like the gentle power of water which is stronger than the massive rigidity of mountains, compassion must conquer the fears and the hatred which set us against each other. We must learn to put aside unfriendly boundaries of nations and hostile dogmas of religious denominations. East and

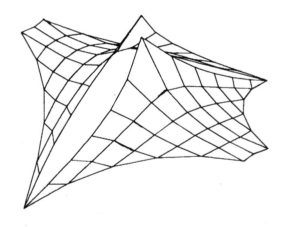

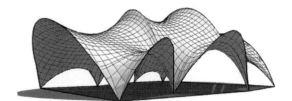

Generating tensile structure surfaces on the computer helps not only the design, engineering analysis, and patterning, but also creates beautiful graphic images.

Above is a shape generated with a formfinding program in the 1980s. It required expert training to run such a program. [1.09]

On the left is a CAD image of a surface shape created with my new design program and transported by dxf. This took just 20 minutes and anyone familiar with computer graphics can learn to do it. [1.10]

Introduction: A Gentler Built Environment

It was a long and hard climb to get to the top of the Denver Airport roof. But the experience was most rewarding and enlightening.
[1.11]

West must meet, North must strengthen South. And we must learn to rebuilt our communities shared by female and male alike, as they were in the distant past when women gave society the shelters in which people lived.

Tensile architecture, with its soft curves, its yielding strength, its glowing translucency, descendent of ancient homes, belongs to such a new era. If it is our determination, that era is in our grasp.

This stick-framed dome structure for a house in Nigeria illustrates a type of building construction that has been used in many parts of the world for as much as half a million years. It is one of architecture's primary structural forms. [2.01]

2

In the Beginning: Domes

This sketch illustrates the most probable construction of one of the earliest house structures found at "Terra Amata" near Nice, France. Each of 20 units housed up to 15 people. The largest measured 15 m by 6 m. [2.02]

Building houses is one of the oldest occupations of the human race. The earliest known evidence of a house structure dates back about 400,000 years. In 1966, during construction of the cliff road to Monte Carlo, traces of some twenty houses were found in a beach cove near Nice, in southern France. These houses had apparently been used for a number of years as a hunting or fishing camp, occupied in late Spring of each year.

The construction of these dwellings - in a camp now called *Terra Amata* - was similar to that of houses still found in use in parts of Africa, Asia and America during this century. Tree branches or saplings were stuck into the ground, close together, to form an oval floor plan. A ridge beam along the center, carried by sturdy posts, formed the central support. The saplings were bent inward until they touched the ridge beam and were then laced to each other, to form arches across the width of the house. It is most likely that horizontal members were added to tie the arches to each other. These gave the structure integrity and strength and helped complete the support frame for the thatch which formed the enclosure surface. Boulders were placed around the periphery to firm up the the ground around the anchorage.

The study of surviving ancient communities shows that variations of this type of stick-framed dome structure were used in many parts of the world and followed traditions developed over many generations. The ground

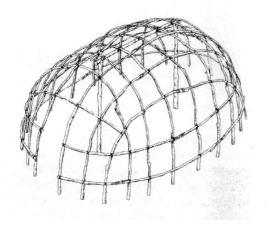

The orthogonal stick-framed dome structure of this North American wigwam resembles the structures found in Terra Amata and in many ancient communities of Africa and Asia. [2.03]

This photograph shows women thatching a grass house at the Trans-Mississippi International Exhibition at Omaha, Nebraska in 1898. .[2.04],

plan of these houses was round or oval. In one tradition the arches follow a *radial* arrangement, all meeting at a central peak like the spokes of a wheel; in another they intersect each other at right angles to form an *orthogonal* grid.

For thatching, natural and readily available materials such as palm or banana tree leaves, grass, straw or reed were used. Woven grass mats or reed panels speeded up the process. Other surface materials were animal skins, bark, reinforced clay, and - much later on - woven textiles. The material selected and the form in which it was applied gave the enclosure its capacity to protect the space against sun, rain, snow and wind. It kept the inside cool in the summer, warm in the winter, and allowed for ventilation.

Building a house was usually a community effort. Although in some traditions the women were the sole builders and owners of houses, the process of house construction was most commonly a social event in which both men and women participated. Usually the men would gather and prepare the frame material. Sometimes they would stake out the ground plan. Most often the women would erect the structural frame and do the thatching, while the men made ritual jokes about their slowness, perhaps having to do with the fear common in

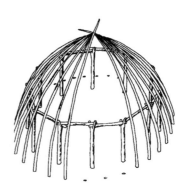
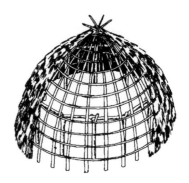

These sketches illustrate stick-framed domes with radial geometry as used in grass houses common in some American Indian communities. [2.05]

In the Beginning: Domes

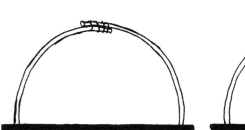 

A continuous arch form is achieved by tying two flexible sticks parallel to each other. It appears later in architecture as the semi-circular or Romanesque arch. A pointed arch form is created by letting the two segments cross each other at the top, not unlike the structural form characteristic of the Gothic cathedral. [2.06]

These South African houses are covered with grass matting. The radial patterns are quite regular with an artistically detailed frieze over the entrance. [2.07]

This house in Dalaba, Guinea, uses layers of straw for thatching material. The "tiled" thatching pattern is the basis for roof tile construction still in use today. [2.08]

many communities, that the house would not be completed before the end of the day; to leave it open over night would invite evil spirits.

The form of these stick-framed dome houses derived from the process of construction and the handling of naturally available materials. Yet its geometry of intersecting arches and hoops resulted in an ingenious structural system of great efficiency because it carried the forces from its own weight and from wind and weather down to the ground in the most direct way.

If we measure structural efficiency by the ratio of the weight of the structure to the external load (wind or snow) which it can carry, the stick-framed domes of the ancient communities are among the most efficient structures ever built. When covered with thatch or matting these domes could not possibly have weighed more than a pound for each square foot of surface area (5 kg/sq.m), yet they were capable of carrying loads many times their own weight. In fact, it is unlikely that they ever failed due to the impact of the forces to which they were subjected; their life span was limited by other factors, such as the durability of the materials and the changes demanded by the traditions and necessities of the community.

Structures with deep, properly shaped curvatures, such as arches or suspended cables, are vastly more efficient in transmitting loads than shallow structural members such as beams and trusses. The proper shape is one which is in total balance with the loads it carries. Structural engineers call it a *funicular* shape. It is the shape which a structure takes on when it has no stiffness at all, acting as if every point of it was hinged. A chain is such a structure. Shaped by its own weight, a chain is the most familiar example of a funicular form.

The problem with such a flexible structure is that its

shape will change when the loads change. A more stable structural network is created by linking a number of chains into a two-way net. Now they can carry changing loads without drastic changes of shape, because there are many different paths the internal forces can take in the process of bringing the external loads to the support points.

Replacing the chains with solid sticks and turning the entire system upside down produces one of the basic dome configurations, which has a remarkable resemblance to the principal geometric shape of the early stick-framed domes of ancient house construction.

The ancient dwellers who built them followed no such theories. They found their dome forms intuitively. Using their hands, sticking flexible saplings into the ground and bending them towards each other, led them to invent the arch. Depending on the way they connected the two halves to each other at the center a pointed or a continuous arch were created. A combination of arches tied together with hoop-like horizontal ties resulted in the elementary dome shapes which form the structural frames of these early houses.

The two basic forms of the ancient stick-framed domes - *radial* structures with the arches all coming to one central hub, and *orthogonal* structures with two sets of parallel arches intersecting each other at right angles - are still the two primary geometric configurations for large-span domes in modern engineering.

The success of these ground-supported domes was based not only on their structural efficiency and the simplicity of their construction process. There were other critical aspects. Their envelope, integrating wall and roof into one continuous surface shape, resulted in a *"minimal surface"*, as today's structural engineers would call it. It enclosed a maximum of usable space with a minimum of surface area, reducing the amount of material needed and the time to assemble it and maintain it. It also conserved the energy needed, such as firewood needed to keep it comfortable in the winter nights.

In areas with no suitable plant materials other solutions were required for enclosing the building. Clay, reinforced with straw fibers, was one of the most successful answers. It made the building's exterior skin waterproof and airtight; and it provided insulation.

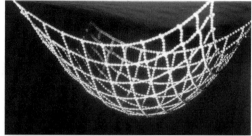
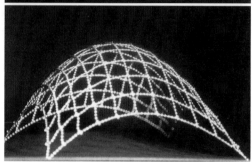

The shape of the intersecting chains in the tensile structure at the tope results from the force flow which keeps them in balance with their own weight. By turning the tensile structure upside-down and substituting chains with rigid members, the grid dome below it is created. It is one of the most efficient structures for spanning space. [2.09],[2.10]

Mat frame housing in Nigeria. The stick-framed dome is covered with colorful woven matting. This evolved obviously much later when herds and weaving were part of the pattern of life.[2.11]

In the Beginning: Domes 13

These houses in a village of northern Cameroon were built entirely of clay. The exterior pattern was not just decoration, but also serves as steps to climb the house for building and repairing it. Each visible ring is one day's work. [2.12]

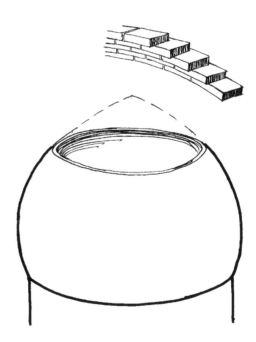

Circular domes were construct- ed without formwork by adding brick rings of varying diameter on top of each other. [2.13]

Working with clay led eventually to the use of sun-baked bricks, which in some areas replaced the stick-built dome. This first manufactured building material was created by forming brick-shaped components and exposing them to the sun to harden.

Skilled craftsmen built domes of various sizes and shapes without the need of formwork. They shaped circular domes by adding one ring of bricks on top of another reducing the ring diameter as they approached the top. The result was the pointed dome which is characteristic in many areas of the world. This simple construction method made it possible to built domes with large spans.

For shallow domes over rectangular or oval spaces, the process was adapted to move in a horizontal rather than a vertical direction. Starting from one edge, one vertical arch segment is added to the one before, overlapping the bricks in the staggered pattern known from wall construction. By using mortars such as gypsum, the brick units adhere sufficiently for the mason to stand on the arch sections just completed, while laying the bricks of the next one.

The fact that buildings formed with these small solid components could not be moved suggests that they belonged to a later period when people had settled into more permanent communities. Entire towns were built using what we now call *"adobe"* construction, and many beautiful domes of impressive size were constructed. I had the opportunity of observing craftsmen at work

Ice house in the Kerman region of Iran. As the strong horizontal rings show, this large dome was constructed using the method illustrated on the previous page. [2.14]

installing such shallow vaults in Teheran, Iran. They moved forward with amazing speed and precision, covering large areas in a manner competitive with modern mechanized construction methods.

The invention of fired bricks came, no doubt, from experience with adobe house construction. Even then houses had fireplaces for cooking and heating, and it is quite plausible that the interior of a house caught fire, developing enough heat to turn the clay into fired brick material. Building a kiln to make modular fired brick units was the next logical step. Magnificent brick domes were built using these dome construction methods. They were often finished both inside and outside with brilliantly colored ceramic tiles. Among those best known are the Taj Mahal in India and the mosques of the Middle East.

Of the great urban civilizations, Rome was the first to match the need for large covered assembly spaces with the technical capacity to produce them. Domes were the only structural system known to serve the purpose. The first large dome covers the Pantheon, built 1900 years ago. Interestingly, it was not built in brick or stone, the most prevalent construction materials of the Roman empire, but rather in a new structural material: concrete. Deep coffers form the underside of the dome, created by the concrete formwork with the purpose of reducing mass. These coffers create the image of a grid of arched ribs and horizontal rings, echoing the outlines of the principal structural elements of the ancient stick-built domes.

Three centuries passed before another dome of this magnitude was built: the *Hagia Sophia*, or "Holy

The dome of the Royal Mosque in Isfahan is one of the most beautiful structures built with rings of brick and covered with glazed tiles. [2.15]

In the Beginning: Domes

The Pantheon is the oldest large-span covered space. It was completed in A.D. 127. Its concrete dome has a diameter of 43.4 m (142.5 ft.) [2.16]

Mimar Sinan's Selimiye Mosque in Edirne, Turkey, completed in 1575, is among the finest of his many domed buildings. [2.17]

Wisdom", was the powerful symbol of the adoption of Christianity by the Roman Empire. It remained a singular achievement for 900 years until Renaissance architect *Filippo Brunelleschi* produced the dome for Florence Cathedral (completed 1436), followed a century later by *Michelangelo's* St. Peters Cathedral in Rome, (completed 1535).

The only place were large scale masonry domes were built in large numbers was in Istanbul and other parts of Sultan Süleyman's Ottoman Empire by his architect *Mimar Sinan*. To built domes in stone masonry was an enormously difficult task, especially when the vaulting techniques were adapted from the linear arch rather than employing the simple methods which adobe craftsman had perfected over thousands of years. These masonry domes required the knowledge to work out the geometry for each piece of masonry and the skill to shape it precisely. Formwork was needed to erect the dome because the masonry would only be capable of carrying its own weight after the last piece - the *keystone* - was in place. And special equipment was required to lift the heavy masonry units onto the formwork.

It took the combination of exceptional motivation, political and economic power, and technical skill to overcome these obstacles. Long before Sinan's days this combination came together to build the hundreds of magnificent vaulted stone cathedrals of the Christian Middle Ages. The Church had the power and the money; the skill was provided by the "masons", a guild of highly trained professionals who understood the material they worked with and knew how to shape it into powerful structures. They were educated to visualize spatial geometry, and they possessed great artistic sensitivity, and enormous patience. Their ambition was not large span but great height, and to achieve it in a way that would make massive stone look light and soaring. The delight of these elegantly carved spaces is the play of daylight streaming in through the enormous stain glass windows - the television screens of the time, often depicting stories from the Bible or events in the history of the Church, informing and entertaining the congregation.

It is the industrial age, however, with its much larger spans, which returns to the pure original dome shapes of the ancient communities, making full use of their superior structural efficiency. Engaging the immense strength of new materials such as cast iron, wrought iron, steel and concrete, unprecedented spans could be achieved. The train sheds and exhibition halls of the 19th century are early examples. Many of these large-span structures exhibit a new feature: starting with the *Crystal Palace* in 1855, glass was used as the new enclosure material. Since the slender, high strength structural members caused little visual obstruction, the interior space was lit with bright daylight. Structure became light, elegant, soaring. Enclosure almost vanished.

Most dome structures were parts of spheres with the main ribs arranged along radial lines. The Astrodome in Houston, Texas, completed in 1968, was the largest translucent structure of this type with an outer diameter of 218 m (714 ft.). It was the first roof structure to cover a full size stadium with a football field and 52,900 spectator seats.

Many new structural forms were explored in search of construction efficiency for large span systems. Civil engineers, engaged in designing bridges, dams, water tanks and reservoirs often took the lead by applying scientific methods to the design and analysis of structural systems. The use of these new structural forms as part of architectural designs often came decades later. Just a few of the more interesting ones are touched on here.

Concrete shells, abandoned with the fall of the Roman Empire, were newly discovered. Felix Candela's shell structures were formed along hyperbolic or saddle shapes which could be generated from sets of straight

The Astrodome under construction in 1965. The "lamella" grid of trussed steel elements spans a circle of 218 m (714 ft.) diameter. It was covered with translucent acrylic panels. [2.18]

Felix Candela's hyperboloid concrete shell structure for a restaurant in Xochimilco, Mexico, 1958. The concrete is only 10 cm (4in.) thick. Its strength depends entirely on its curvature. [2.19]

In the Beginning: Domes

Pier Luigi Nervi's Hangar in Orbitello, Italy, completed in 1944, was made of precast concrete elements. The structural behavior of this daring design was investigated by model testing. [2.20]

lines. Pier Luigi Nervi built his delicate, daring domes from precast concrete elements of exceptionally slender dimensions. He developed a new technology, called *Ferro Cemento*, which resulted in concrete construction so light that even ships could be constructed with concrete. Heinz Isler most recently used experimental and digital methods for finding the most efficient concrete shell shapes requiring the least amount of materials; his shells are only 7 cm (3 in.) thick and their organic forms are of breathtaking elegance.

In spite of these advances, I believe that concrete shells have not found wider use because they face the same obstacles which hampered the masonry domes: the difficulty of their design, the need for formwork, and the opaque surface they produce. The clear, beautiful lines of the computer graphics and of the formwork which Heinz Isler uses in the design and construction of his shells are again reminiscent of the ancient stick-framed domes, hinting in the direction of new solutions

The German architect and pioneer of tensile architecture, *Frei Otto*, used an orthogonal grid dome geometry for his design of a *grid-shell* structure at the Garden Exhibition in Mannheim, Germany, in 1968. This ingen-

This dazzlingly beautiful shell by Heinz Isler is shaped by form-finding methods like those used for tensile structures. Therefore, the entire shell is under compression. Most of the 7 cm thick concrete shell requires no flexure reinforcement and is waterproof, because there are no cracks. [2.21]

ious design, in which Otto was assisted by the British engineer Ted Happold, uses sets of thin, continuous wood struts covered by a translucent fabric. It demonstrates that spaces of great geometric variety can be created starting with simple square grids.

A variety of very beautiful grid shell structures haven been built in the last few years in Japan, England and Germany. Jörg Schlaich has developed an especially intelligent system: he introduces prestressing tendons as thin diagonal members. This makes the grid a true shell in the form of a Schwedler Dome.

The structural system which in the 1970s replaced rigid dome structures for full size stadium covers is also based on a geometry of intersecting structural lines, bowing upwards like arches. It was derived from experience with the air-supported fabric structures which had been pioneered by Walter Bird and had been successfully used for radar covers, warehouses, and tennis court covers.

By giving the surface a low profile curvature and reinforcing the fabric with a grid of high strength steel cables, large and super-large spans could be achieved at a fraction of the cost and construction time required for conventional structural systems. The geometric configuration developed by the late David Geiger, then my partner in the consulting firm of Geiger Berger Associates, has an uncanny kinship with the ancient orthogonal stick-framed dome shapes with their sets of parallel arches and their oval ground plan.

But here the resemblance ends. The oval shape of air-supported domes has a precise mathematical definition. Its outline is a super-ellipse, belonging to a family of curves ranging between ellipse and rectangle. The elevations of the intersection points of the cable grid are computed to generate internal equilibrium at all intersection points of the cable net and, at the same time, to load the peripheral compression ring so that it is *funicular* under uniform load. This means that the ring carries load in the most direct, efficient way. The shapes of the ring and the cable net as well as the patterns of the fabric pieces are determined by computer, as is the prediction of the structure's behavior under wind, snow and other external loads.

What was intuitively conceived and approximately realized in the ancient stick-domes is now scientifically shaped and precisely constructed. The weight of the

Grid dome structure at the Mannheim Garden Exhibition in 1968, designed by architect Frei Otto with engineer Ted Happold. The irregular, organic form is generated from a square grid of continuous wood struts covered with translucent fabric. [2.22]

In recent years new grid dome systems have been developed, especially in Japan and Germany. The photo above shows the Lehrter Bahnhof in Berlin, Germany, designed with the elegant system by Jörg Schlaich. Note the diagonal prestressing tendons and the exposed cables for correction of the funicular shape of the arches. [2.23]

In the Beginning: Domes

The air-supported roof of the U.S.Pavilion at Expo 70 in Osaka, Japan, the first low profile long span structure of its kind, which was conceived by David Geiger. [2.24]

At right is the typical plan geometry with a super-elliptic rung. [2.25]

Above: Unidome air-supported roof before re-construction. [2.26].

Below: Grid dome system with trussed arches and cable rein - forcement. [2.27]. Below right: The completed grid dome with central fabric roof, 1998. [2.28]

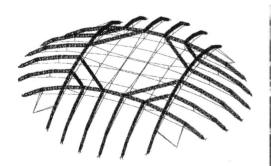

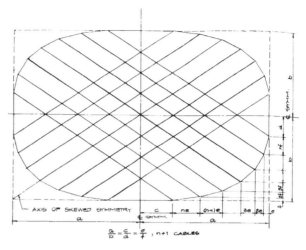

Mathematically the shape of a super-ellipse is expressed by the equation $(x/a)^m + (y/b)^m = 1$
For an ellipse $m = 2$, for a rectangle $m = \infty$
With $a=b=r$ it is a circle or a square, respectively.
For a typical dome, $m \sim 2.3$.

translucent air-supported dome - though capable of spanning 20 times as far - is almost exactly the same: 1 lb. per square foot (5 kg/sq.m).

Many of the eight major stadiums with air-supported roofs are still in use worldwide. Though their dependence on mechanical devices has proven to have its problems and has led to a number of sometimes expensive and disturbing deflations, the success of the air-supported domes has led to a greater acceptance of fabric structures, opened the way to new, less controversial structural systems of similar efficiency, and made the use of daylight a highly desirable feature of architecture.

The replacement of the Unidome's air-supported roof by a grid-dome in 1999 is an example which demonstrates the power of these historic forms when translated into contemporary technology.

Arab Black Tents use the most advanced tensile technology of the pre-industrial age. They can be erected or taken down in a matter of minutes [3.01].

The rectangular panels of the Black Tents can be folded and loaded on a camels back. [3.02]

3
From Tents to Tensile Architecture

Transportable Mogul yurts like these were most probably used to move the housing of Genghis Khan and his armies. [3.03]

Although it usually took only a day to build a traditional stick-framed house, for Nomadic people who needed to move quickly in pursuit of their livelihood and take their dwellings with them, this was too long. They adopted structures with lighter building components, which were more transportable and could be erected anywhere in not much more than an hour. The Tuaregs of the Western Sahara, for instance, reduced the composition of their stick-framed domes to a minimum number of arches. They covered their frames with mats or with animal skins which could be rolled or folded into small bundles. Similar movable dwelling systems were common in many parts of the world. Among the best known are the yurts used by the Turks, Tartars, and Mongolians. Yurts were the housing units used by Genghis Khan and his troops as they conquered large parts of Europe and Asia.

Assembling a Tuareg frame tent. A minimum number of members make up the frame. Foldable skins or mats are used for the cover. [3.04]

Although tipis were light, easy to put up and take down, they were comfortable housing units which could adapt to various weather conditions. The tipi could be closed up or opened, flaps could be adjusted to control wind and air flow, and a protective fence could be installed around the periphery in the winter. Insulation could be added between double skins and a fire could be lit. [3.05]

The tipi of the American Indians served in a similar way: although heavier, their frame members were straight. For transport they were bundled. The lighter end of each pole of two bundles were attached to either side of a draught animal. The other end was simply dragged along the ground behind. Historically, dogs were used for the purpose. In more recent times the dogs were replaced by horses, permitting the use of much larger tipis.

The construction of a tipi starts with laying out the ground plan. It is an egg-shaped oval with the smaller radius at the entrance side which is commonly oriented towards the rising sun. The back of the tipi faces the prevailing winds with the aero-dynamic shape of the uneven oval acting to reduce the forces from air flow around the structure.

Depending on the tradition, the erection of the frame begins with setting up a three-pole or a four-pole frame. In both cases the poles are laid on the ground and tied to each other near the top. Two poles, forming an A-frame, are pulled up in position with the help of the one or the two additional poles, setting up the basic structural frame. By adjusting the position of their ground support points, the

Crow woman transporting tipi poles and covers. Bundles of poles are attached to a horse, skins are draped over its back. [3.06]

Tents to Tensile Architecture

location of the center point is perfected and the alignment of the poles refined. For better wind resistance the cone is always kept steeper at the rear.

With a firm central support now established at the top, the rest of the poles can then be erected simply by laying their upper ends into the V-shaped crotches, and placing them in the proper position at the ground.

The traditional enclosure was animal skins. (The use of woven canvass is a very recent development). The entire enclosure was made of one piece, which had the form of a semi-circle. It required anywhere from eight to more than twenty buffalo skins, which were tanned and smoked to keep them from becoming brittle. The skin was wrapped around the conical frame so that the splice would occur at the entrance. Flaps at the bottom would function as doors, others at the top as ventilation units and wind guides. Various positions of the door and ventilation flaps would help in controlling the comfort in the tipi. Around the periphery the outer skin was attached to the ground with stakes to secure the light structure against turning over in wind storms. This attachment was easily adjustable, so that the edge could be turned up

Women of the Blackfeet tribe building a tipi, ca. 1900
[3.07]

for better ventilation.

A liner covered the inside to a point five or six feet above the ground, serving as thermal insulation and sometimes as decorative interior finish. With the outside surface folded up at the ground, fresh air would flow between the two skins, exhausting at the top. This was an effective way of cooling the interior on hot summer days. In cold winter periods the space between the two skins would be filled with dry grass to improve the insulation. Snow would be banked around the base on the outside.

The roles of men and women in the construction and ownership of the dwellings were clearly defined. In most Indian communities of central North America women were in charge of the home. They built or supervised the building of the tipis, and they owned them. A divorced man had to return to his parents to have a roof over his head. In other areas men and women shared the construction, with the men most commonly responsible for preparing and erecting the poles and the women making and applying the skins.

On the outside the surface was quite often decorated. This was a task most commonly given to men with special training.

It was still tedious to transport housing units with cumbersome and heavy frame members and erect them at each stop, especially on trips which took more than a day. Still lighter and more easily transportable shelters were needed by Nomadic people who had to travel often

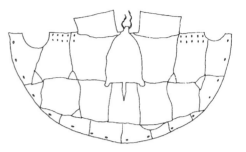

Patterning of buffalo hide cover. [3.09]

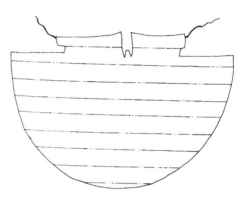

Patterning of canvas cover. Canvas replaced buffalo hides late in the 19th century. [3.10]

Sioux tipi cover with decorations. The large suns, here on the east and west walls, are feathered, because the Sioux believed the sun to be a bird which flies from east to west. [3.08]

Tents to Tensile Architecture

Reconstruction of tent from campsite discovered in Moldova, Russia, dated 40,000 B.C. Animal bones and tusks were found which had been used for poles. Earth berms and boulders forming the perimeter of the tents were still there. [3.11]

and cover large distances. The answer was found in reducing the size and number of rigid compression members to a minimum, replacing them, wherever possible with flexible lightweight tensile components. The result was the tent.

Because tents are made entirely of biodegradable materials, it is hard to determine just how long they have been with us. We find the oldest evidence of their existence in the cold climates of the north. The people of Siberia, Lapland, Iceland, and Alaska, who were compelled to follow the animals they hunted and had no natural construction materials, used animal skins suspended from poles made of large bones. The remnants of a campsite in Moldova, Russia, have been dated at approximately 40,000 B.C.

Desert people, such as the Bedouins, Moors, Berbers, Baludas and Kurds, developed tent structures which they could load onto their camels - or even their horses - and which they could erect anywhere in a matter of minutes rather than hours or days. The construction of *black tents*, for example, shows all the important features which make tensile structures work then and today: A woven fabric cover, patterned and assembled from rectangular strips, is draped over ropes which, in turn, are supported by poles located in the center and along the edges. Anchor ropes transfer the loads from the

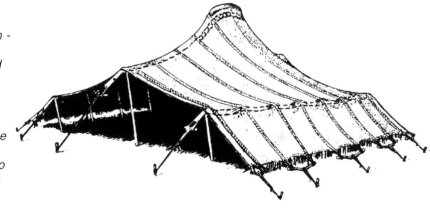

Arab black tents show the most advanced tensile technology of the pre-industrial age. A fabric skin is draped over ropes along the ridge and the eaves. Struts form supports along these rope lines. Stakes anchor the ropes and the fabric into the ground. The materials are foldable and light enough to be carried on the back of a camel or even a horse. [3.12]

An A-frame forms the central support of this Moroccan tent. A saddle shaped strap picks up the load of the woolen fabric. The corners are held by posts, tie ropes and stakes. [3.13]

fabric panels to the stakes which hold the structure down and outward.

In its simplest form - the tents of southwest Morocco - the fabric forms a square pyramid supported by one central A-frame. The A-frame consists of two sticks which cross each other near the top and support an arched saddle piece. Four ropes define the corner ridges of the four-sided tent. They are anchored into the ground by wooden stakes. Some of the corners are held in vertical position by short poles.

As in today's technologically advanced tensile structures, the fabric is put under stress to keep it from flapping in the wind. Introducing internal stress is part of the assembly and erection process. The process begins by laying out the ropes and fabric panels and attaching them to each other. Stakes are put in the ground and the tent edges are attached to the stakes. Then the poles are pushed up into position, giving the structure its shape and putting it under stress.

It takes little to turn this ancient tent into a contemporary tensile structure: substituting wool with a modern non-combustible fabric and giving it a scientifically computed pattern; replacing the ropes with galvanized steel cables; making the A - frames and posts from structural steel or aluminum; putting concrete anchors in the ground; and, finally, stressing the structure to a predetermined stress level.

Tents remained important for many purposes throughout history. But having settled down in fixed communities, relying on agriculture and trade for food, the great societies of "historic" times built durable buildings for their homes. Tents were used for secondary purposes. They still provided shelter for the great armies of

Throughout history tents were used for portable housing, especially by armies. This illustration of Charlemagne's camp is by Tavernier Renant de Montauban in "Croniques et Conquestes de Charlemagne". [3.14]

Tents to Tensile Architecture

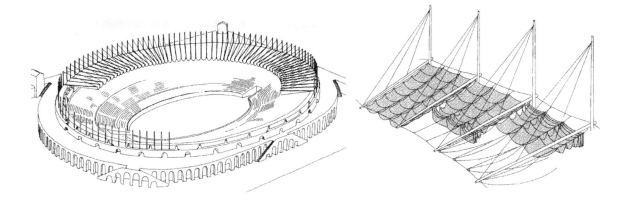

Reconstructions by Rainer Graefe of Roman shade structures, called "vela", common in arena buildings. At left is a drawing of the amphitheater in Pompeii. The detail at right shows the configuration of the vela system. This technology was derived from sail ship rigging and was operated by retired sailors. [3.15]

The use of sophisticated tensile technology in sail ships is ancient. The rigging of this chinese junk has been used for millennia. [3.16]

the Persians, Greeks, and Romans. They served this purpose throughout the Middle Ages, the American Civil War and the wars of this century. Indeed they are still critical for modern armies.

Tensile structures were also used for religious events, social gatherings, and shade structures. The center of public worship of the Jewish people after the Exodus from Egypt was the Tabernacle, a large square enclosed by fabric structure walls.

Among the more sophisticated applications of tensile architecture in this category were retractable shade structures installed in Roman amphitheaters, providing protection against the hot sun for the spectators who assembled in these spectacular entertainment centers.

We know that the Romans did not have the technology - or the motivation - to enclose the Colosseum or any of the many other amphitheaters with a tensile structure roof. They did, however, have sophisticated fabric shade structures called "vela". Learning from sail ship technology, retractable fabric panels were suspended from horizontal "masts", which cantilevered from the back wall by a system of ropes and struts not unlike the stay systems which stabilizes the masts of sail ships. Masonry details, including special buttresses, stone extensions with circular holes, and stones with square cut-outs in exterior walls show where struts were anchored, and allow reconstruction of their configuration. The vela are clearly the forerunners of tensile structure systems now used to cover the seating sections of modern stadiums.

Not only was the design of the vela system derived from sail ship technology, but retired sailors operated these retractable shade roof installations. Indeed, sail ships provide an unparalleled source of experience with fabric tensile forms and technologies. Sails take on many shapes. Ropes or cables, stays and masts, and the combinations of configurations in which these components are arranged appear in great variety through the long history of naval architecture.

The evolution of modern tensile structures began with the construction of large-span suspension bridges in the early eighteen hundreds. Though their designers had the advantage of a new scientific understanding of force flow in structures, they applied the same suspension principles which had been used for thousands of years in the rope bridges of the Far East and South America.

The efficiency of a rope to carry load across a span is obvious. Anyone who has hung laundry from a clothes line understands it. Being flexible, the rope changes its shape to fit each load configuration so it will always carry load in the most direct way, namely in a simple line of tension. A draped rope or cable loaded in one point only will take on a triangular shape. Two loads will make it trapezoidal, many loads polygonal. Under a continuous, uniform load (such as its own weight) its shape will approximate a parabola.

A suspension cable shaped by its own sustained load distribution is called a *catenary*. Catenary cables look like upside-down arches. But since cables are loaded in tension and arches are loaded in compression, much thinner members can be used. This is so because, under load, thin compression members (arches, columns) have two ways of becoming shorter. They can compress along their central axis, or they can bow out

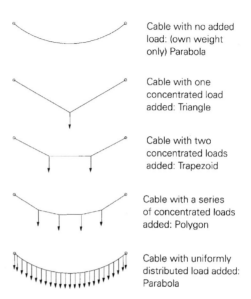

Shapes of a single cable due to various load patterns. [3.17]

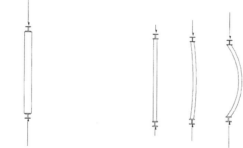

Tension and compression.[3.18]

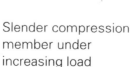

Thick compression member under increasing load

Slender compression member under increasing load

Tension member under increasing load

Tents to Tensile Architecture

The Kuanksien An-lan suspension bridge in western China, shown at right, is carried by bamboo cables. It has eight spans of up to 66 m. It is more than 2300 years old. It is closed for two months of every years for renovation. [3.19]

The Ching-Lung wrought iron chain bridge over the Yangtse river, shown above, spans 100 m. It is estimated that Chinese wrought iron chain bridges of this type go back 1900 years. [3.20]

sideways and collapse, a process engineers call *buckling*. Buckling depends on the stiffness of the structural element. Therefore compression members need to have a certain thickness to carry load without buckling.

Tension members do not have this problem; they have only one way of getting longer: namely by stretching along own their axis. Thickness plays no role. As a result, they can be as thin as their tensile capacity allows. Higher strength enable tensile cables to be thinner, but he thickness of a slender arch or a column in compression is not reduced by higher strength at all because in this case load capacity is limited by its size and elasticity.

As far as we know, the earliest use of this principle for the purpose of crossing a mountain gorge or a river was in the Himalayas more than 4000 years ago. Single ropes, made of bamboo strands, were draped across and anchored on both ends. A person or an animal would then slide across, hanging from a greased bamboo cradle. Bamboo ropes were also used for more important bridges. Rows of bamboo ropes would be hung parallel to each other and covered with a wooden deck. The railings would also be made form bamboo ropes. Bridges of this kind are found in southwestern China dating back more than 2000 years. The advantage of the use of bamboo is its easy availability and the high strength: bamboo is about three times as strong as a hemp rope. It suffers from lack of durability, however, necessitating the annual renovation of such bridges. The invention of wrought iron in China, probably as early as 100 A.D., made it possible to build bridges using wrought iron chains. Spans of up to 138 meters were built which required very little maintenance.

The invention of wrought iron in Europe, about 1700 years later, made the first large-span suspension bridges possible. One of the earliest of these bridges with a span of 177 m (580 ft.) is the bridge over the

Menai Strait in North Wales. It was designed by Thomas Telford. As a skilled mason he had worked on a masonry arch bridges before becoming an architectural draftsman. When offered a position as *county surveyor* he switched from house architecture to public works projects and became one of the first members of the *British Society of Civil Engineers.* The formation of *civil engineering* as distinguished from *architecture* and from *military engineering* signified the beginning of the division of the profession responsible for the design and maintenance of the built environment which has prevailed ever since.

Telford''s introduction of iron bridges was a major step in an evolution which changed construction forever. The Menai Bridge used suspension chains made of

The Brooklyn Bridge over the East River is John Roebling's masterpiece. It uses galvanized high strength steel cables. The granite towers are load bearing. Completed in 1883 its 523 m (1715 ft.) span was the longest in the world. [3.22]

The Clifton suspension bridge over the Avon river gorge, uses three layers of eye-bar chains made of wrought iron . The bars are linked with iron pins. Designed by Brunel in 1832, the 300 m (980ft) span was completed in 1864. [3.21]

wrought iron bars linked to each other by pins. This type of eye-bar chain remained in use for much of the 19th century.

The invention of high strength steel cables revolutionized the construction of suspension bridges. The new cables had ten times the strength of wrought iron. They were composed of multiple, continuous strands of thin, high strength wire and they could be manufactured and shipped in coils of tremendous length. Continuous suspensions members over the full length of the bridge were now possible, eliminating hundreds of pinned linkage points, each of them a potential point of failure of the chain, while breakage of one cable strand by itself was insignificant for the safety of the bridge.

John Roebling, inventor and producer of steel cables, ingenious engineer, and gifted artist, designed a number of elegant suspension bridges in the United States. His masterpiece, the Brooklyn Bridge, is one of the finest bridges ever built. From the superb, load bearing granite towers to the radical arrangement of stay cables, every component is purposeful and economical. Looking at the structure you understand how it works. Its image is based on revealing the simplicity of its structure and on showing the harmony of its parts, and the delicacy of its proportions. There is nothing to be added, nothing to be taken away. It is just right, as a structure, as a building, as a piece of art.

The next big leap in suspension bridge design again took place in New York City with the construction of Otmar Amann's George Washington Bridge over the Hudson River. Its 3500 ft. span was almost double that of the Brooklyn Bridge. The delicate grid of its steel towers was initially intended to be covered by a granite facing. The generous sweep of the main cables in the huge center span is in dramatic contrast to their tight sags over the short end spans, illustrating the simple relationship of force and geometry.

Numerous large span suspension bridges followed the opening of the George Washington Bridge in 1929. Stronger steel materials and more advanced design methods led to lighter bridge designs and with them returned the problem which had haunted suspension bridges throughout their history: lack of dynamic stability under transverse winds. Prehistoric bridges swayed in the wind and had to be closed during storms. The early designs of the 19th century had to be reinforced and

Otmar Amann's George Washington Bridge over the Hudson in New York, completed in 1929, almost doubled the span of the Brooklyn Bridge. [3.23]

stiffened time and again to reduce vibrations. As one of many, in 1854 the 331m (1000 ft) span bridge in Wheeling, Pennsylvania collapsed. It was the largest span bridge of its time, and it was only five years old.

Roebling, who knew of the Wheeling disaster but was not able to analyze the causes scientifically, handled the problem intuitively by adding stay cables and stiffening the bridge decks. Most other designers simply ignored it until 1940 when the Tacoma Bridge collapsed in a steady cross wind of only 18.8 m/s (41 miles/hour) velocity. A few years before, David Steinman had predicted the possibility of this event. He understood the mechanism by which the alternating vortices of the wind currents could build up the sway of the bridge and destroy it when their rhythm matched the structure's natural frequency. Tacoma was on top of Steinman's list. But no one listened.

Miraculously there was only one car on the bridge, driven by a photographer. He abandoned his car, ran off the bridge, set up his movie camera, and took the now famous 25 minute film strip of the collapse of the bridge. After this dramatic event, bridges were designed and strengthened to avoid its recurrence mainly by stiffening the roadway with heavy trusses. The engineers of the Humber Bridge in England found a more elegant solution: its bridge deck is a slender welded steel box designed like an upside-down aircraft wing. As a result, cross winds always cause downward loads, for which the bridge has been designed to begin with. There are no vibration effects due to air currents. The result is an exquisitely slender profile expressing the simple elegance of its engineering concept. Several bridges have since been built using this principle for the bridge deck. The largest and most beautiful of them is the Storebelt bridge in Denmark.

In the 1950's, Fritz Leonhardt began to use stay cables as primary components for large span bridges. His asymmetric bridge design for Cologne, Germany, with a single, A-shaped pylon is still one of the most expressive bridge structures. Stay cable bridges

In 1940 the Tacoma Narrows Bridge collapsed when dynamic excitation caused by cross winds caused it to sway out of control. [3.24]

The Storebelt bridge in Denmark, completed in 1998, has the second largest span in the world. With concrete towers and abutments and a slender welded steel box-girder it is the most elegant of them all. [3.25]

Tents to Tensile Architecture

33

Severin Bridge over the Rhine in Cologne, Germany. The bridge deck hangs from stay cables which radiate from the top of the single A-frame. 1959. [3.26]

are now increasingly used worldwide both for steel and concrete bridges.

Linear tensile structures are also used for other long span applications such as transmission towers and pipe bridges. Some designs for pipe bridges create three-dimensional stability by a combination of suspension and stay cables carrying load in all three directions. For some reason this simple principle has not been used for suspension bridge carrying roadways.

The success of tensile structure designs in civil engineering applications had little impact on building design. One reason, obviously, is that they are most effective for large and super-large spans. Such spans are rarely

Suspension cables in three planes support pipe bridge over the Elbe river in the Check Republic. [3.27]

A cable suspended roof gives this paper mill in Mantua, Italy, the column free space it needs for its production process. Pier Luigi Nervi used steel cables hung from concrete frames to achieve the 148 m (485 ft.) long space. [3.28]

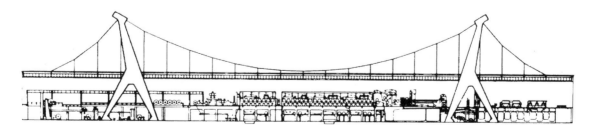

required for buildings. Engineer Pier Luigi Nervi in his design for a paper mill in Mantua, Italy, applied the suspension bridge concept literally. As a result, it looks more like a bridge than a building. Architect Eero Saarinen in his superb design for a the Dulles airport terminal used the simple shape of a hanging cable as its cross section and expanded it into a large draped sheet of concrete hung from powerful columns, higher on one side than the other. The exquisite design of these

"Imaginary Sketch" by architect Erich Mendelsohn. 1917. [3.29]

Eero Saarinen's design for the terminal building at Dulles airport in Washington, D.C has amazing similarity to Mendelsohn's sketch. The draped one-way concrete roof hangs between two rows of massive concrete piers. The roof concrete is put under compression by posttensioning with high strength steel strands. [3.30]

columns as they penetrate the roof and engage its curled up edge, emphasizes the flow of forces and gives the building its distinct character.

In a circular building the tension forces from the cables can be equilibrated by a compression ring, avoiding the enormous cost of Dulles type cantilever columns. Engineer Lev Zetlin's bicycle wheel roof designs made use of this concept. Two layers of radially oriented para-

Engineer Lev Zetlin's cable supported roof over the auditorium of the city of Utica consists of a circular array of opposing parabolic cables which are held apart by struts. This "bicycle wheel" roof system is suited for carrying upward and downward loads. A circular compression ring resists the tensile forces of the cables. [3.31] [3.32]

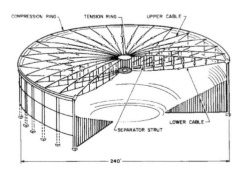

Tents to Tensile Architecture

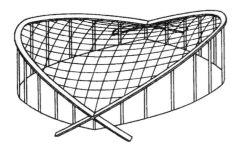

The saddle roof of the Raleigh Arena consists of two intersecting sets of cables with parabolic curvatures spanning between inclined concrete arches as shown in the sketch [3.33]. After more than forty years the building is in excellent shape. [3.34]

The roof of Eero Saarinen's Yale Hockey Rink consists of two saddle surfaces spanning between three arches. Here the cable nets are covered with wood decking. Structural Engineer: Severud Associates. Completed 1958. [3.35]

bolic cables, held apart by steel struts, provide the mechanisms to carry upward and downward loads.

The roof structure of the Raleigh Arena in North Carolina is ingenious in its simplicity. Its surface is shaped by two sets of parabolic cables which intersect each other at right angles. One curves upward, one downward. Two inclined arches form the boundaries of the saddle surface created by the cable geometry. Their weight and that of the roof rest on the mullions of the

exterior glass wall. In their own plane of curvature, they balance the tensile forces from the cable nets which, because of their opposite curvatures, can carry upward and downward loads with equal efficiency.

The shape of the space created by this structure fits the varying height of the sloping seating perfectly. This may, in fact, have been a major reason for architect Matthew Nowicki to explore this concept. The final design was the result of a team effort with structural engineer Fred Severud. When completed in 1953 it was the first tensile surface structure of significance. In its wake saddle shapes were built in many parts of the world to cover large assembly spaces.

There were two problems with the Raleigh design. One was nominal: it emitted occasional screeching noises. The surface, consisting of folded steel deck panels, was stiffer than the cables. Their connectors would slide along the cables to accommodate the incompatible strains. The other was more serious. The light, highly elastic roof, experienced vibrating movements. This problem was solved by adding thin stay wires along the periphery.

Today, after more than 40 years of use, the Raleigh

The Congress Hall in Berlin was designed with a wood covered cable net roof, but converted by German engineer-contractors into a concrete shell. [3.36]

Arena is in excellent shape. Nowicki died much too young in a plane crash. Severud's office became responsible for many subsequent surface tensile structures, including the Yale Hockey Rink, Madison Square Garden in New York, and the original design of the Berlin Congress Hall. It also became the training ground for many designers of tensile structures; I was with them for eight years.

There were many other tensile structure designs with opaque roofs designed and built in the 1950s and 1960s, two decades which showed a distinct appetite for exploring structural forms in architecture before the waves of postmodernism sent "structure" into hiding.

Among them are two of the most daring, most interesting and most beautiful structural buildings of that period: the two Olympic National Stadiums in Tokyo, completed in 1964. Although architect Kenzo Tange derived his forms from traditional tents, their relation to the Yale Hockey Rink is obvious. Both buildings house circular arena seating structures to which entrance halls are appended. Their spiral tensile roof shapes enclose this sequence of spaces perfectly.

The large stadium seats 15,000 spectators for swimming and skating events. Two splayed main suspension cables span between concrete masts which are tied back by extension of the main cables. The main support cable of the 4,000 seat basketball arena winds around its single mast in a dramatic spiral. The outer edge in both buildings is formed by a concrete ring beam.

The roof surfaces are not quite what they seem: instead of being soft membranes on flexible cables which take on their own natural shape, they are semi-rigid steel shells consisting of steel decking welded to

Kenzo Tange's Olympic National Stadiums in Tokyo use tensile forms for the brilliant architectural designs. The spiral design of the basketball arena is shown in drawings. [3.37; 3.38]

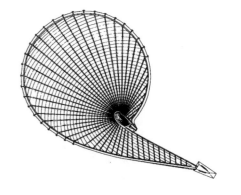

The roof of the swim stadium spans between splayed ridge cables and the sweeping curves of the edge beams. Both structures can be seen in this aerial view. [3.39]

Frei Otto's superbly grafted tents at the 1964 Lausanne Exhibition illustrate the grace of pure tensile forms. [3.40]

rigid, preformed steel girders. To avoid vibration problems due to the shallow curvature in some areas of these roof surfaces, anti-vibration devices were built into the design.

Fabric tensile structures began with architect Frei Otto in Germany. From the outset he understood the indispensable principles of this building technology: that structural and architectural form are inseparable; that flexibility is strength not weakness; that the surface material must be more pliable than the elements which support it.

As a fighter pilot, his life had depended on the reliability of a flexible light weight structure. As a student of architecture among the ruins of post-war Berlin he saw the impermanence of "permanent" buildings and understood the necessity of getting things done with minimal means. Early on he began to explore and investigate the nature of tensile structures with an amazing level of dedication, beginning with saddle shape model studies in the early 1950s. He rediscovered the tent as an architectural building form. And he studied every variation of tensile structure form imaginable.

What makes his work so appealing and so significant is the aesthetic quality of every sketch, every model,

Workers on the cable net of the German Pavilion at the World's Fair in Montreal begin to raise the fabric skin, illustrating the extreme lightness of the cable net structure. [3.41]

every detail of a structure he produced. His intense involvement with technology was always an integral part of his existence and development as an artist.

The sensuous, almost spiritual lines of his early tents caught the imagination of the architectural profession worldwide. Tensile structures became acceptable, even desirable forms of architecture. One can assume that without this effect of Frei Otto's work none of us would be building them today.

The German Pavilion at the World's Fair in Montreal, Canada, gave him the first opportunity to build a large structure. Although its free-form cable net was a thing of great beauty, it had its technical problems: deviating from his earlier structures, the tensile surface was no longer a membrane but a cable net. This made its construction more difficult. The secondary fabric skin

The completed pavilion is an elegant and powerful free-form sculpture which exhibits most of the features which Frei Otto developed as natural expressions of the flow of tensile forces. [3.42]

Tents to Tensile Architecture

hung below it - an arrangement that made it vulnerable to the accumulation of snow which would be dammed by the cables. In fact the structure, only five years old, failed in a snow storm.

Frei Otto's masterpiece is the the roof of the Munich Olympic stadium, built for the 1972 Olympic games. A powerful catenary forms its interior edge. Flying masts ride on suspension cables reaching down from the tall masts and running on to the catenary. They help create high points which generate the vaulted volumes of the undulating ceiling which soars over this great space. Despite the interfering geometry of the rigid acrylic panel cover (forced upon the designers by televisions extravagant daylight requirements), these powerful tensile forms make the structure one of the great landmarks of the twentieth century.

The roof of the Munich Olympic Stadium is one of the great architectural structures of this century. Despite the unfortunate pattern of the rigid acrylic panel cover, the soaring tensile forms dominate the space. [3.43]

Computer generated surface shape of a multi-unit radial tent structure. [4.01]

4

How Tensile Structures Work

The cable supported roof of the Madison Square Garden arena in New York under construction. [4.02]

Conventional building structures made of concrete, steel, wood, or masonry have two main properties which make them stable and capable of transmitting load: gravity and rigidity. Masonry walls stand up because of their bulky weight. Steel frames cary load by their stiff strength of resisting bending. In tensile structures gravity and rigidity are not available as the critical structural properties. Fabric structures, in particular, are so light that their weight is almost negligible. And the materials of which they are made, such as fabric and cables, are highly flexible. Other means have to be harnessed to give stability and strength to a structural system consisting of flexible members. Their components must be arranged in a specific geometric form (*surface shape*) and must be subjected to a specific pattern of internal stresses (*prestress pattern*).

The geometry of tensile structures is, therefore, not arbitrary but follows strict engineering rules. Once the boundaries and support points of the structure have been set and the prestress pattern is selected, there is only one three-dimensional surface shape under which the structure is in equilibrium at all points. The exact configuration of this surface shape is not known beforehand but has to be found in a mathematical process called *form find - ing* or *shape generation*.

Fortunately, the basic principles which control the relationship between prestress pattern and surface shape are relatively simple. The actual computation of the geometry is, of course, performed by computer. (This process is described in some detail in Chapter 8). The purpose of this chapter is to develop an understanding of the major forms which satisfy these conditions.

We have already discussed why tensile members are more efficient structural elements than compression members. Slender compression members tend to avoid carrying load by *buckling*. Unless they have a certain minimum thickness they bend out of shape and fail independent of their direct material strength. They bow out, one could say. Structures made of tensile members, on the other hand, become more stable with increasing stress levels. The stress will pull each member into line, making the structure taut. Thinner tensile members are better in adapting their shape to the load pattern. Making full use of the material's high strength they will carry the load most directly in uniform, concentric tension. The load acts along the center line of a rope or cable. Each fiber of the rope or each strand of the cable carries its even share of stress.

There is, however, one particular problem which comes with this flexibility: one single tensile member by itself will not make a reliable structure because it will change its shape drastically when the load changes in magnitude, direction or distribution. A clothesline with laundry takes on a smooth downward bow in response to its own weight and the weight of the pieces of laundry which hang from it. We call this funicular shape of a cable a *catenary*. As long as there are no other forces acting on the system, it retains this shape. If, however, we add a wet shirt in the middle, this shape will change, deflecting downward where we added the shirt, and flattening out in other places to retain its overall length and stay taut. If there is wind, the situation changes dra-

Clothesline: No wind. [4.03]

In a gust the clothesline loses its shape. [4.04]

A steady wind flow may even reverse the curvature. [4.05]

How Tensile Structures Work

Weights could be added to hold the line down [4.06]

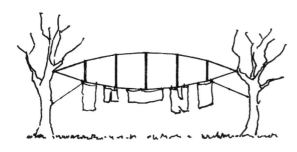

A second cable of reversed shape could be added above to retain the shape and resists upward loads. Stays are needed to keep it from rotating. [4.07]

A cable added below is the more efficient solution for stabilizing the clothes-line. [4.08]

matically: under gusty conditions, the clothesline will lose its initial shape, flapping up and down. Steady upward wind may even reverse its curvature for a time.

During this process our line will go through a "slack" stage during which the stabilizing tensile force is lost. This chain of events may be acceptable for a clothesline where - at worst, some laundry may fall off and get dirty - but for a building structure such uncontrolled behavior is not acceptable. Steps have to be taken to control the behavior of tensile structures and their components.

The problem of the clothesline could be solved in a number of ways. Weights could be hung on the line. This would be awkward and require a stronger rope, but for a building it might be a useful idea. The roof of the Madison Square Garden arena in New York, for instance, carries the mechanical equipment room, acting as the "weights" to hold the structure down.

Alternatively, we could stabilize the clothes- line by adding a reversed cable on top, pushing down on the line with struts. This by itself would prove ineffective as the structure would simply turn itself sideways or upside down. Stay cables anchored back to the tree will avoid this. (In the circular arrangement at the Utica arena, which we saw earlier, such "bow string" cable trusses work perfectly without special stabilizers).

A reversed rope line below the main rope with vertical strings connecting the two lines is

Section of cable roof, Madison Square Garden. [4.09]

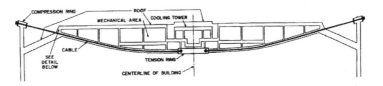

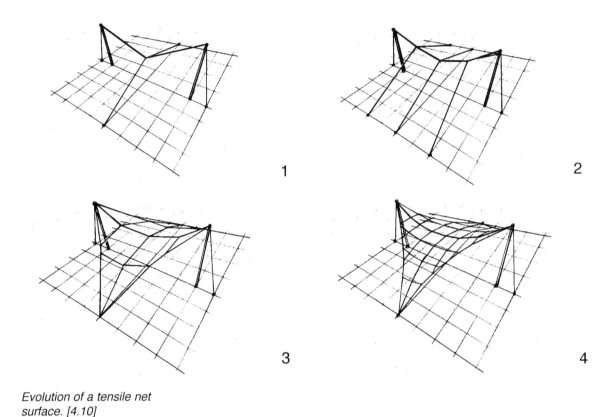

Evolution of a tensile net surface. [4.10]

Force diagram at a typical node in a tensile saddle surface, and part of of a typical anti-clastic surface net. [4.11]

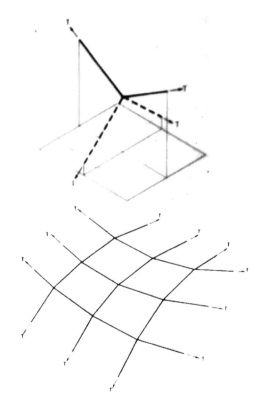

simpler and more effective. This is the most stable and the most sensible solution.

If we now consider these two lines to be cables in a roof structure and turn the lower cable by 90° to run across the upper one, we achieve a stable point at the place of their intersection [1]. Adding two more tie-down cables parallel to the first one generates two more stable intersection points [2]. The addition of edge cables placed between the upper and lower support points begins to turn this arrangement into a two-way cable net [3] which is completed by adding denser sets of cables, parallel to the original, ones in each direction

How Tensile Structures Work

Four point structure built by my students at the Technion in Haifa, winter 1998. The membrane is made of Polyethylene. [4.12].

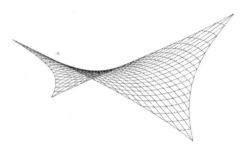

Stress line diagram of a four point structure. [4.13]

The shape of a simple four point structure can be produced by stretching a square piece of flexible material out of its flat plane. A piece of panty hose may work best. [4.14]

We have now arrived at a two-way cable net structure which has one unique characteristic: any two cables which meet at an intersection point, or *node*, are bent in opposite directions, one pulling downward and the other pulling upward, thereby exerting pressure against each other. When this cable net is put into tension we increase the internal pressure at each node. The higher the initial tension, or *prestress level*, the more stable and stiff is the cable net system.

The surface structure we have generated by this process, is called a *four point structure*. It is the simplest form of a saddle-shape, a basic structural geometry which we encountered in the roofs of the Raleigh Arena and the Yale University Hockey Rink.

You can easily produce your own model of this basic surface shape: just take a cloth napkin (or even better, a piece of panty hose or stretch fabric) and lay it on a table. Hold down three corners with a finger at each corner. (This is a social game: you need at least two people as it only works if there is a little tension put into it!). Then pull up the fourth corner, and you have the basic shape of a four point structure.

The edges could be four rigid beams. A better solution is to use four edge catenaries. To make it a real structure the three low points need to be firmly anchored against uplift and lateral movement. The high point needs a vertical and horizontal support, such as a mast and a cable.

We now have all the components required to make a

strong and stable structural system. If we proportion the elements correctly we can carry snow and wind at any level of load for which the structure needs to be designed. Snow will principally be carried by cables which sag downward; wind uplift by those curved upward. In either case the edge catenaries will transmit the cable forces to the anchor points, which, in turn, will bring them safely down into the firm ground.

Prestress is introduced by the simple means of pulling one corner outward. Shortening one of the tie-down cables will do this. It has the effect of tilting the mast outward. The edge cables connected to it want to flatten out. This in turn puts stress on the net cables which all interact with each other. Provided that all members of the cable net have been fabricated to the right length and are connected to each other in the right places, all members will end up with the exact tension they were designed for. One common choice is for the tension to be uniform in the entire net.

If the surface is not a cable net but a continuous fabric membrane, the net lines shown represent strips of fabric in the structural analysis. For patterning purposes the net lines - in one direction only - represent edges of fabric strips of which the membrane is made. Cable lines still represent cables. Edge catenaries are running in a fabric sleeves or are attached to the membrane with metal clamps.

This simple form is beautifully illustrated by the four-point structure which was built by my students at the Technion in Haifa, where I taught for a semester in the fall of 1998. The membrane is polyethylene sheeting which costs just a few cents per square foot. In my designs I like to slope the masts and keep the cables vertical. This makes the anchorages easier and keeps people from running into sloping cables. (If they run into a mast I can't really help them!)

As all this demonstrates, *four* is the absolute minimum number of anchor points for a tensile structure. A surface generated by connecting only three points will be a flat triangle. Since curvature is a critical aspect of a tensile structure, four

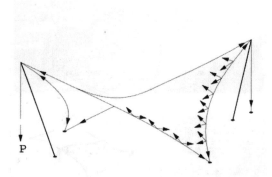

Force flow as a result of pulling with prestress force P at the left tie-down. [4.15]

support points are needed, with one of them in a different plane than the other three.

This phenomenon distinguishes tensile structures from rigid structural systems which require three support points for stability. A stool with three legs is perfectly stable; a table with four legs may wobble, since only three legs are needed for a particular load distribution and the fourth leg might not touch the floor, at at any rate not without inserting a matchbook under it. So we see that for static, rigid structures, the dominant geometric element is the triangle, the critical number is three.

For objects and creatures which move, the number three is not sufficient. My granddaughter had a dog who had lost a leg in an accident. He somehow managed to hobble around, looking almost as awkward as humans when they walk. (Having only two legs, humans move by leaning forward until they are about to fall on their faces. In order to prevent this potentially embarrassing event they quickly extend a leg forward; and having now gained a step in the process, repeat it).

A normally constructed dog has four legs: three to stand on, and one to move forward with. Equally, a well conceived automobile has four wheels: three to roll on and one to pop in and out of potholes. Without that fourth wheel - as anyone knows who has driven a three-wheeler - the

How Tensile Structures Work

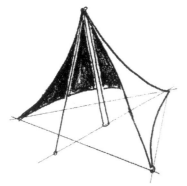

Single four point structure. Three points share the ground as common plane. The fourth is created by the pole. [4.16]

Two four point structures held by a single pole cover a square. [4.17]

Four four point structures cover a larger square. One radial cable could replace each pair of adjacent edge catenaries in the slope of the structure. [4.18]

Weidlinger's tensegrity structure of the Georgia Dome in Atlanta is covered with diamond shaped fabric panels. Construction photo, 1993 [4.19]

undampened shock caused by one of the three wheels crashing into the hole is both mind- and teeth-shattering.

For dynamic structures, therefore, the critical number is four and the ability for the fourth support to flex - the spring in the wheel of the automobile, the muscles and tendons in the dog's leg - is essential.

In this sense, *tensile structures* are dynamic structures. Their critical number is four, and the flexibility of their elastic components help them adjust to changing load conditions. The dominating geometric element is a four-sided trapeze. The four-point-structure is its simplest manifestation.

The geometry of four-point structures can have numerous variations depending on the positioning of the support points. Combining two or more four-point structures leads to a large variety of interesting tent forms. Several four-point structures can be combined by sharing their high points. Larger spaces can be covered by combining a number of four-point structures and varying the height of their supports. As the Georgia Dome in Atlanta demonstrates, they can also become modular elements in surfaces covering very large spaces.

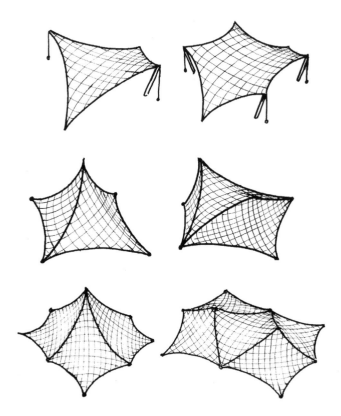

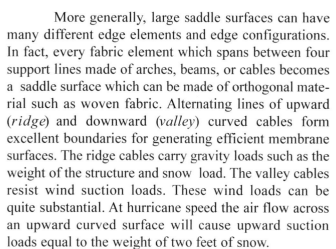

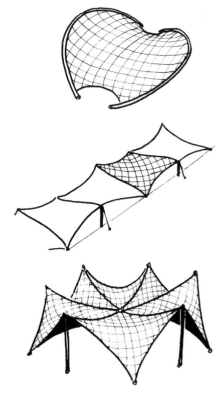

Variations and combinations of saddle surfaces. [4.20]

More generally, large saddle surfaces can have many different edge elements and edge configurations. In fact, every fabric element which spans between four support lines made of arches, beams, or cables becomes a saddle surface which can be made of orthogonal material such as woven fabric. Alternating lines of upward (*ridge*) and downward (*valley*) curved cables form excellent boundaries for generating efficient membrane surfaces. The ridge cables carry gravity loads such as the weight of the structure and snow load. The valley cables resist wind suction loads. These wind loads can be quite substantial. At hurricane speed the air flow across an upward curved surface will cause upward suction loads equal to the weight of two feet of snow.

The space between these alternating cables is filled with a surface representing a double curved net consisting of lines parallel and orthogonal to the cables. If the apex of the valley cables is located a sufficient distance below the bottom of the ridge cables the fabric membrane represented by this net can carry both the downward and upward loads to the cables at safe stress levels.

The simplest structural configuration which this alternating cable pattern can generate is one which is

Model for Folk Life Pavilion in Philadelphia. Saddle surfaces span between two rows of mast. 1976. [4.21]

How Tensile Structures Work

At the Denver airport terminal, saddle surface liner panels span between ridge and valley cables. [4.23]

suspended between two rows of masts. In this case the ridge cables hang between the tops of the masts, the valley cables are anchored to ground anchors located on the same line as the masts and half way between them. To prevent the mast tops from moving inward, tie-down cables are required which are anchored outside the structure.

This arrangement has some disadvantages. The forces from the tie cables will greatly increase the load on the masts. Also, the high points of the structure are at the outside face, creating large openings which, in the case of an enclosed building, creates large end walls. Moving the masts inward greatly improves this condition. The shape of the ridge cable then looks exactly like a suspension bridge and the shape of the valley cable like an inverted arch. The terminal roof at the new Denver International Airport is designed this way.

There are, of course, many variations to these basic surface shapes. Applications of quite a number of them are shown in the structures which are described in Chapter 6.

An interesting case is the arrangement of alternating ridge and valley cables in a circular configuration. An important advantage of a circular or oval shape is the fact that the external forces need not be tied down to the ground but can be gathered into a ring beam. Closed systems, meaning structures whose geometry permits to short-circuit the force flow by use of funicular ring beams or ring cables, often have a substantial economic advantage over open systems where all the forces have to be brought to the ground and anchored.

This circular configuration is, in a sense, the transition to radial tent structures. Radial tent shapes are the alternative to saddle shapes. They are the only other way of satisfying the principal requirement that at any node in the net two intersecting lines pull in opposite directions. Surfaces defined in this way are called *anti-clastic* surfaces, which means that the curvatures in the two major axes are of opposite direction. All tensile structures must have anti-clastic shapes. (*Syn-clastic* surfaces, on the other hand, are illustrated by intersecting net lines which curve in the same direction. Domes and pressurized membranes are typical examples of structures with synclastic surface characteristics).

The two principal anti-clastic surface geometries derive directly from the coordinate systems which are used to generate them. If we start with a square grid (Cartesian coordinate system with x-y-z coordinates oriented at right angles to each other), we arrive at the saddle shape. In its purest form it consists of two sets of intersecting parabolas. Lines oriented diagonally to the parabolas are straight lines.

If we start with a radial grid (Polar coordinate system with f-r-z coordinates), we generate a radial tent. It consists of a set of lines radiating from one point which intersect with another set of lines which form concentric rings. In a radial tent structure the two sets of net lines are quite different from each other. The radial net lines, originating from a single point, spread out towards the edge. The ring lines forming complete hoops have no edge at all.

The fact that the ring lines are self-contained is a main advantage of the radial tent structure. It requires no costly end connections in this direction. One end of the radial net lines goes directly to one common point. Only the other edge of the radials needs to be connected to another member - a beam, an arch, a cable - for further transmission of its end forces to fixed supports.

A disadvantage of the radial tent is the triangular shape of the fabric panels. This is a shape which tends to lead to more complex patterning, more seams and less effective utilization of fabric material which is manufactured in long, rectangular roles, the dimensions of which are determined by the weaving process. A further disadvantage is the need for radial cables to bring the load to the top connection point.

An array of multiple tent units can be used to cover

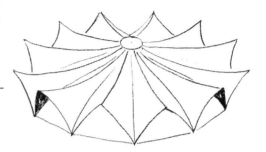

Fabric surfaces spanning between ridges and valleys in a rotational arrangement. [4.24]

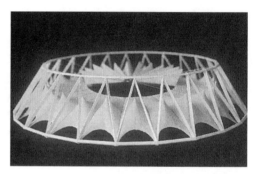

Arena roof design using external triangulated ring structure for a closed system design. Tent units span between ridges and valleys. The center is retractable. 1978. [4.25]

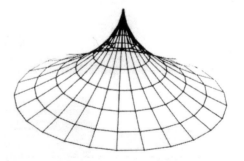

Net of primary radial tent shape with radial and ring lines. [4.26]

How Tensile Structures Work

Radial tent units of Jeddah Haj Terminal under con- struction in 1980 [4.27]

large roof areas. The world's largest roof structure, the Haj Terminal of the Jeddah International Airport in Saudi Arabia, consists of 210 radial tent units arranged in ten modules. Each unit with a square side length of 45 m (150 ft) has 32 radial cables and is framed by varying combinations of valley cables and edge catenaries. This arrangement requires three different tent modules: for the interior, the edge, and the corner. A design which rotates the units by 45° makes all tent units become identical. Triangular edge units are then needed if the roof plan is a rectangle as is shown in the first illustration of this chapter. (Again, a number of radial tent designs are illustrated later in the book).

Giving the structure the right shape is, no doubt, the most important aspect of the design. Stressing it so that it can hold this shape without excessive deformation and without going slack in critical areas, is second in importance. Making certain, that no part of the structure will fail or be damaged under load is third. Non-linear analysis methods are required to check the design for these last two aspects. Chapter 8 describes this process in some detail.

At this point let me show what happens to a structure under load in one particular case. The structure is a radial tent with a single mast, such as the "Great Adventure" roof structures described in Chapter 6.1. The edge consists of a rigid steel frame on top of a one story window wall. The section shows the curvature of the radial cables.

A horizontal wind stream moving from the left to the right puts inward pressures on the wind side, and causes outward suction force on the lee side, with the result that the force in the radial cable on the wind side will increase, that on the lee side will decrease. This change in cable forces will result in an unbalanced force at the top of the mast pulling towards the left. The mast will begin to tilt in that direction. This movement will bring the support points of the windward cable closer together, increasing its sag. It will stretch the support points of the leeward cable apart, making it flatter. Since the force level in a cable changes with the inverse of the sag, the force in the windward cable goes down, that in the leeward cable goes up. The mast will tilt until a new equilibrium of forces is reached at its top. In this gentle process the forces change little because the structure flexes. Were the top of the mast held fixed, the force change would be large. " Letting go" is the most important rule in the design of tensile structures.

How Tensile Structures Work

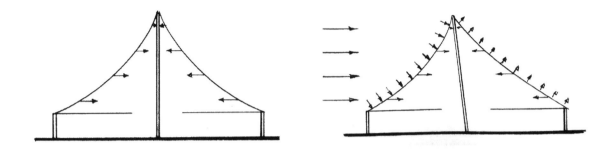

Under lateral wind load the structure moves to minimize the stresses in the system. The diagram at the top shows a radial tent structure under prestress forces only. The diagram below shows the deformed section. The mast has moved into a new equilibrium.[4.28]

5
Materials for Tensile Structures

I am standing on one of the ten modules of the Jeddah Haj Terminal roof in Jeddah, Saudi Arabia. Structural fabric, especially when properly stressed, is easy to walk on. [5.02]

Teflon coated fiberglass fabric, galvanized high-strength steel cables and painted tubular steel masts are the materials which form the roof of Canada Place, Vancouver, B.C. [5.01]

Structural fabric is the material which defines light weight tensile structures. As a primary structural component, structural fabric must have strength to span between supporting elements, carry snow and wind loads, and be safe to walk on. As the enclosure element, it needs to be airtight, waterproof, fire resistant and durable. In most cases, it also needs to transmit daylight, reflect heat, control sound, and be easy to keep clean.

Structural fabrics in common use today consist of a structural base material, such as fiberglass or polyester cloth, covered with surface coatings such as PVC (polyvinylchloride, popularly called vinyl), Teflon or Silicone. The much less expensive PVC-coated polyester fabrics are generally used for temporary structures. Teflon coated fiberglass fabrics have have been used in most permanent building applications completed in the last three decades. Silicone coated fiberglass was used in the 1980s on a number of projects and it is another possibility, as are many other combinations of materials. New versions of PVC-coated polyester with dirt resistant top coats and tear resistant weaves show increased strength and durability at little additional cost. They offer the potential of greatly reducing construction cost and thereby widening the market of fabric structures for permanent buildings.

In the last few years, very low cost fabric materials

have entered the market, such as woven and knitted high-density polyethylene. Although they have sufficient strength, they are (at least for now) deficient in other properties necessary for permanent building applications. However, their emergence suggests that fabric technology is only at a beginning, and an increase in demand for fabric structures in architecture will be met by the technology required.

The structural fabric membrane is often reinforced with or supported by linear flexible elements, usually high strength steel cables.

Finally, a tensile structure needs rigid structural elements to support the flexible fabric and cable membrane, generate its peaks, form its edges, and create the anchors which hold it down.

For the tensile structure to function properly there must be a hierarchy of elasticity of the materials. The membrane must be more flexible than the cables. They, in turn, must stretch more than the rigid members which support them. When such a hierarchy exists, structures are easy to built and behave in a predictable and efficient way under load.

Rigid support elements are masts, struts, frames, arches and edge beams. They are generally made of materials which we now consider "conventional" despite the fact that none of them existed existed even a hundred years ago. These materials include high strength steel, reinforced and prestressed concrete, laminated wood, structural aluminum, and, most recently, a number of composite synthctic materials.

The rigid elements in tensile structures must be strong, light, reliable, readily available, and easy to fabricate, transport and erect. Structural steel satisfies these requirements excellently. The drawback of steel is that it must be protected against corrosion by painting or galvanizing (which requires careful application and maintenance).

Structural steel - a great improvement over wrought iron and cast iron - appeared on the construction market just over a century ago. The much higher strengths of today's steel products and the advances in structural design methods have combined to improve steel construction to the point where a building frame today requires one third of the steel needed fifty years ago. This is critical, because, due to its use in consumer goods such as automobiles and to related permanent losses by oxidation, the supply of iron-based materials is in rapid decline. The brown

The tent units of the Tennessee Pavilion in Knoxville are suspended from painted steel A-frames. [5.03]

dust resulting from this process, commonly called rust, disappears into the earth, never to be recovered.

Aluminum is a prime candidate to replace steel in lightweight structures. It comes with similar strengths, weighs one third, and does not corrode. Reduced costs of aluminum production and advances in welding methods, have made it competitive to steel in many applications, especially for exterior, exposed components. As a natural resource aluminum (Al) is much more abundant than iron (Fe) and little of it is lost to corrosion.

Fiberglass, carbon, and other synthetic materials have begun to appear on the market as alternative structural framing products. They are extremely light and very strong, but they are still expensive and have other shortcomings. In the

Materials for Tensile Structures

Trussed steel arches support the fabric of a tennis roof cover in Atlanta. [5.04]

Laminated wood arches at Bullocks Department Store in San Jose, California, span 96 ft. [5.05]

At the AELTC Indoor Practice Facility at Wimbledon, England, the fabric membrane hangs from precast concrete arches connected to piers which are cast in place. [5.06]

long run, though, synthetics will play an increasing role in the production of structural components.

Structural lumber in the form of laminated wood is suitable for use in arches, frames, or columns. These components are made with advanced fabrication methods which convert strips of wood cut from ordinary trees into high strength structural members. With proper reforesting programs, wood used for structures need not be a cause for depletion of our natural resources.

All of these materials are vulnerable to fire, because under extreme heat they lose their strength. Fire protection can, however, almost always be achieved easily enough and at reasonable cost. Components can, for example, be located a safe distance from the potential source of fire; they can be encased with protective materials; or the adjacent area can be sprinklered.

It should be noted that fabric tensile structures are inherently much less vulnerable to fire damage than most conventional enclosure structures. Fire tests have demonstrated that the seams of most fabric materials will open up and exhaust hot fumes and smoke long before enough heat develops to harm fabric, cables, and other structural components. The continuity of its tensile elements and the extreme light weight of the actual structure will greatly reduce both the risk of injury to people, and damage to the building as a whole. In fact, the most likely result of a major fire in a fabric-covered building is a hole in the roof which can be patched relatively easily and quickly.

Concern about fire and weather protection do not

apply to concrete. Concrete is a fireproof material; when properly designed and built, it should outlast the other parts of the building by a huge margin. Although concrete was discovered and extensively used by the Romans, as an advanced structural material it is entirely a product of the 20th century. You could call it a synthetic stone made of natural soil materials which include cement as the binder that holds it together. By adding steel reinforcement to overcome its tendency to fail under tension, concrete is transformed into the powerful structural material we use today in the construction of almost all of our buildings.

Concrete, is easily available in most parts of the world, can be shaped into many forms. As a construction material it is inexpensive, especially when shaped properly and when used to produce surfaces (floors, walls, roofs) which serve both as structure and enclosure of a building. Because concrete components are heavy, they are best concentrated around the periphery of tensile structures, or used as its direct supports.

Cables serve a number of functions in tensile structure applications: reinforcement of the fabric where the spans and stresses get too large; linear tension support elements along ridges, valleys and edges; tie-backs and stays to stabilize rigid support elements; and "redundancy" members which provide alternate routes for the flow of forces and assure the integrity of the overall structural system should a fabric panel have to be replaced. The predominant cable material is high strength bridge strand, a product which was developed for suspension bridge construction in the mid-nineteenth century.

Bridge strand cables are made of bundles of parallel high strength steel wires which arc twisted slightly so that under tension the wire strands press against each other. The high strength of the wire strands is achieved in various ways, one of which is to draw them through a succession of dies of decreasing diameter. Strand strengths in the order of 270 000 lbs/sq.in. (1.87 million kPa) are achieved, which is five times the strength of structural steel sections common in building construction.

Cables made of these strands have enormous tensile load capacity. They can be fabricated to almost any desired length. They are flexible enough to be coiled on large, transportable drums. They are provided with reliable, standardized end fittings which make it easy to

Hefty concrete triangulated concrete piers, reminiscent of the flying buttresses of medieval cathedrals, support the San Diego Convention Center roof. [5.07]

A common type of High Strength Steel Cable: Bridge Strand. This cable has 37 strands.[5.08]

Materials for Tensile Structures 59

End fittings of main stay cables at Canada Place, Vancouver.[5.09]

connect them to each other or to rigid structural support components. They are fabricated to precise tolerances and can be proof tested in the shop to levels above any load they will experience in the building. They can be galvanized to protect them against corrosion and are available with shop-applied plastic sheathing which gives them a long life-span even when they are placed outside the building skin and thereby exposed to the natural elements.

Although there are new alternative materials with highly desirable properties - Kevlar, fiberglass, polypropylene, and carbon - nevertheless, high strength steel cables have so far remained the most reliable and economical material for linear tensile members.

The most important component of a tensile structure is the fabric membrane. It is the key to the successful functioning of a tensile structure. Although excellent materials have been developed and successfully used for several decades, the future of tensile architecture still depends on the evolution of fabrics which cost less, are easier to handle, and have better properties for their use in permanent buildings than the materials in use today. Three particular aspects need to be discussed to evaluate structural fabrics: structural strength, behavior during construction and in use, and surface properties.

The ability of the fabric to carry load depends mainly on two properties: tensile strength and tear strength. Grab a sheet of writing paper in the middle of the two short sides and try to pull it apart. If you do it carefully you most likely will not be able to break the paper. Now hold the paper in two spots close together along the same edge, using the thumbs and forefingers of both hands. It takes no effort at all to tear the paper in two. Similarly, fabric will not easily fail in tension but is vulnerable to

It is hard to break a sheet of paper in tension, but it's easy to tear it. The same is true for fabric. [5.10]

tear.

Tear failures start at an open edge or at a hole in the fabric. It is, therefore, critical that fabric panels be contained all around the edges and that this continuity is meticulously maintained. Most commonly this is achieved by edge ropes in continuous sleeves which are connected to cables or other structural members by clamping devices. Almost all failures I have experienced were tear failures which started from a point where a bare cut edge provided the opportunity for the start of a tear. All but one of these fabric tears occurred during construction.

The only tensile failure I experienced, happened during the construction of the first module of the Haj Airport project in Jeddah, Saudi Arabia, and it was very dramatic! A review of the background of this failure is a good path to understanding critical properties of structural fabric, especially since it led to the most thorough experimental study of structural fabric. This study was carried out in the Granville, Ohio, laboratories of Owens Corning, the contractor of the Haj roof. To understand this failure we first need to review the basic construction of structural fabric materials and look at some of the of the essential results of the study.

The base material of the fabric used for the Haj project is woven fiberglass. In the typical basket weave used for this fabric the warp (or woof) threads are initially straight and the fill (or weft) threads snake around them in alternate patterns as the result of a weaving process in which alternate warp threads are pulled upwards or downward and a fill thread is shuttled between them. In the resulting long rolls of fabric the fill threads, running side to side, are kinked around the straight warp threads which run the full length. The coating process, in which the Teflon coating is applied in five passes, holds this configuration in position.

Coated material in this state is used to fabricate structural membranes based on computer-generated patterns. During construction these membranes are being stressed to a pre-determined level (prestress level). When bi-axial stress is applied to the fabric the stressing process will kink the warp threads and reduce the kinks in the fill threads until the pressure exerted at each crossing of threads is in equilibrium. At the same time, this added stress stretches the fibers in both directions.

In a typical case of uniform bi-axial prestress - the

At the Bullocks department store in San Jose, CA, edge catenaries placed inside continuous fabric sleeves transmit the stress of the fabric to the anchor points. [5.11]

Materials for Tensile Structures

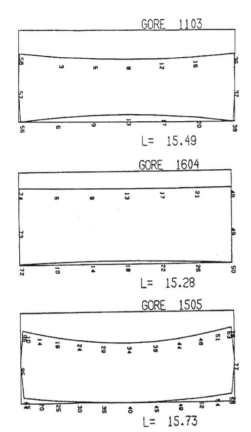

Computer generated patterning layouts. The rectangle indicates the uncut fabric role. [5.12]

Jeddah structures, for example - the elastic stretch of the fibers in both directions might be 3%. In the fill direction the straightening of the kinks might add 3% of elongation. In the warp direction the added kinks might take away 3% of the length of the treads. As a result, the fabric extends by 6% in the fill direction, while in the warp direction it does not change its length at all. This is not an unusual condition in actual structures.

In order to compensate for this change in lengths the material has to be fabricated shorter or longer by an amount called its compensation. The compensation in the fill and warp directions are usually different. In our case the fill compensation is 6% and the warp compensation is 0%. The compensations for each roll of material are different and are determined by a number of biaxial tests. The fill compensations in Jeddah ranged between 3% and 10%. These are substantial numbers. They mean that in the case of a 6% fill compensation a membrane with a fill dimension of 100 ft.(30 m) has to be fabricated 6 ft. (1.8 m) short and, during the stressing process, has to be elongated by 6 ft. (1.8 m).

This behavior of fabric was well known before the Granville testing started. Information on the fabrics strength under load, especially under load sustained for a long time, was not very precise. Here the investigation produced the most valuable information.

The simplest and most basic measurement of a fabric's capacity to bear load is its *strip tensile strength*. The "strength" of a material is defined as the stress recorded in a clearly defined, standardized test specimen at the moment of failure. The strip tensile strength is measured on half-inch wide strips of fabric pulled in a testing machine where the load is applied in a few seconds.

While for most materials the strength is not time dependent, for woven fiberglass of the type used in Teflon-coated fabric the strip tensile strength varies greatly with the length of time a test sample is kept under load. The tests showed that a fabric which fails under instantaneous loading at 800 lbs may fail at 650 lbs if the load is sustained for half an hour. This sustained load capacity will further decrease for several months, eventually reaching a permanent value in the order of 450 lbs or not much more than half the strength it carried under instantaneous loading.

The tests also showed a substantial difference between "virgin" fabric samples and "damaged" material which had been deliberately folded, creased and mangled in an attempt to simulate the conditions the fabric may be exposed to during transportation and erection. The strength of the damaged fabric could be reduced to half the strength of the virgin material. In combination, fabric as used in an actual project and subjected to sustained load could have a strip tensile strength of one quarter of the strip tensile strength. The test samples of 800 lb. material could fail at 200 lb. In the actual building, due to biaxial stress conditions, a failure could occur at even lower levels of stress. To account for such influences and make up for variations in material quality, errors in evaluating loads and other factors, structural engineers use a safety factor of at least 1.5 to arrive at a safe allowable stress. For the example above this allowable stress would be 200/1.5 = 133 lbs/in.

With this knowledge we go back to the airport construction site in Jeddah in February of 1980. During the stressing process of Module A one of the 21 tent units called unit P clearly behaved different than any of the other units. Module A was the first one of ten modules, each with plan dimensions of 1050 ft. (320 m) by 450 ft.(137 m), covering an area larger than the roof of the Astrodome. Walking over unit P felt like walking on concrete, while the other units felt spring like a trampoline which is typical for fabric tensile surfaces at prestress level. With the help of hand-held testing devices

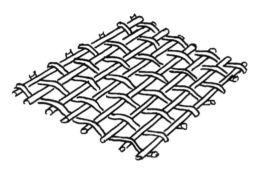

*Warp and fill configuration before stressing.
The warp is straight, the fill weaves around it. [5.14]*

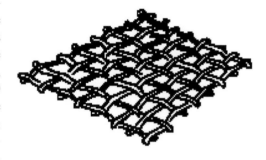

Warp and fill configuration after stressing, At equal stress the kinks are equal. [5.15]

Overstress due to wrong compensation caused these four identical rips in unit P of the Jeddah Haj terminal roof during construction. [5.13]

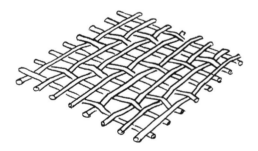

Warp and fill configuration after stressing of a 4H satin weave. [5.16]

Materials for Tensile Structures

designed to measure fabric stress we found extraordinary stress levels near the corners of unit P in areas which are least influenced by the jack located at the peak. In fact these areas showed stresses of over 150 lbs./in. or about three times the stress level expected from the jacking forces and confirmed by stress readings in other fabric areas. There was no explanation for this condition.

The stressing process was monitored from a central electronic board from which electric jacks at the peaks of the units could be operated one at a time or in groups of any desired combination. They would pull the top rings of the tent units upward, stretching the membranes and thereby stressing them. The system allowed sophisticated adjustments to the stress patterns of the module. Although I had no notion as to the cause of the problem I had the contractor try out various changes in jacking force patterns. Nothing worked.

At the end of the day, with none of the maneuvers having any success and still with no idea as to the cause of the problem, we decided to reduce all jacking forces by 20% and close up for the night.

Sipping a scotch-on-the-rocks - Craig Fallon, the immensely capable construction superintendent, had managed to procure a bottle of Scotch, no easy job in Saudi Arabia - we uneasily decided to sleep on the problem and see what we could do in the morning. It was only two hours later that loud banging at the door of my trailer woke me from a deep sleep. "She ripped", somebody yelled.

When I got to the site ten minutes later with a few associates I had collected in the rattling old Chevy at our disposal, we could see the stars through a gaping rip in the fabric stretching from the top ring down into one of the corners of unit P. By next morning all four corners had ripped in the same fashion, and by 10 a.m. I was in the hot seat at a huge conference table, confronting high level Saudi government officials, trying to convince them that this was not necessarily the end of this $180 million project, that the other twenty units were o.k., that this unit would be replaced within days, that......

Why did it happen, they wanted to know.

It took half a year of diligent detective work to find the answer to this question. What had happened was that unit P had been cut wrong. The compensation had been based on tests from a different role of material. It was about 3% short in the fill direction. In the corners where the the fabric is almost as flat as a drum with little room for re-adjustment of shape, this substantial error in length led to the 150 lbs/in readings. There were undoubtedly spots with still higher stresses. This was a sustained load condition. And the

Owens Corning developed a testing machine which measures biaxial stresses to full failure levels.
[5.17]

huge tent unit had been folded several ways to fit it into the box in which it was shipped which measured 16 ft. x 8 ft. x 4 ft. (5 m x 2.5 m x 1.2 m). So the conditions, discussed before, applied.

With an original tested strip tensile strength of the fabric of about 800 lbs/in this could lead to failure of test specimen near 800 x 1/2 x 1/2 = 200 lbs/in. In the real membrane it could fail at even lower level at the weakest spot. The effect of cooling from the desert heat of the day to the chilly night temperature would raise the fabric stress by another 30 pli. That was enough to break it.

It confirmed the generally accepted rule of setting the safe allowable stresses for permanently sustained loads such as prestress at 1/8 of the tested strip tensile strength of the virgin material. In the Jeddah Airport case this would have meant a maximum allowable prestress level of 100 lbs/in. The actual design prestress level was 60 lbs/in, Prestress levels on more recent project are usually even lower.

For wind loads whose peaks are generated by short term gusts, maximum design stresses of 1/4 of the strip tensile strength would then be safe since there is no sustained load impact. For snow loads whose 50 year peak could last for a few weeks, 1/5 of the strip tensile strength has been adopted as appropriate allowable stress. Since these are the commonly used design parameters, the findings of the Jeddah investigation were confirming accepted practice.

With these basic requirements, which apply to Teflon-coated fiberglass, fabrics with strip tensile strengths between 500 and 600 pli are generally sufficient for the design of tensile structures. These values also apply to the more pliable silicone-coated fiberglass fabrics, which should have smaller losses of strength due to handling, Polyester fabrics experience practically no weakening due to handling. But they undergo losses of strength with age due to the impact of ultraviolet light on the polyester base material. Therefore, the tensile requirements developed for fiberglass fabrics again lead to sensible and safe choices of fabric materials.

The flexibility of the double curved fabric membrane makes expansion joints unnecessary. A continuous skin can cover a building of any size. [5.18]

After the proper prestress has been achieved in a tensile structure, the membrane acts sufficiently like an elastic, isometric structural surface in spite of the basket weave kinks. Modeling it as a two-way net still is the most representative basis for digital analysis. The usefulness of fashionable finite-element methods of analysis is overrated and may even lead to distorted results if not applied with proper understanding.

Because fiberglass fabrics fail abruptly when they reach their tensile strength, safety factors need to be conservative. The softer polyester fibers are much safer in this respect since, long before they brake, they reach a yield point beyond which they stretch extensively.

These are the parameters which make the

Materials for Tensile Structures

For transportation the fabric is folded in multiple layers. Softer fabrics are easier to handle and experience less damage.
[5.19]

The arch supported silicone-coated roof units at the Sibley Horticulture Center at Callaway Gardens transmits 50% light.
[5.20]

use of fabric material structurally reliable. A critical further requirement is a fool-proof seaming method which is compatible with the full strength of the material. There are other characteristics which have an impact on the quality and economy of the structural membrane. Materials with greater widths need fewer seams. Softer materials are easier to handle. Compatibility with gasketing and clamping devices is important. The choice of coating material has a considerable impact on these. It is also critical for the fabrics durability and fire resistance when the structure is part of a permanent building.

There are non-structural properties which are important and often critical in the choice of fabric for a specific building application. They are most often the result of the combination of base fabric and coating.

Translucency is the most important of these, since bringing daylight into the building may result in significant energy savings and in producing a healthier, better functioning and more attractive interior environment. Low heat absorption allows further energy savings, sometimes even the elimination of air-conditioning. Surface properties which avoid the accumulation of dirt or make it easy to remove accumulated dirt, help to maintain not only the sparkling appearance which is now associated with fabric structures but also help retain the translucency and reflectivity levels required for efficient functioning of the building.

Polyester fabrics with PVC coatings, by far the least expensive materials available, may last 15 to 20 years if special precautions are taken to protect the polyester against ultra-violet rays. Fire safety depends on the mechanism of seams opening, since these materials are combustible. Dirt resistance requires special surface finishes.

Teflon-coated fiberglass has become the primary material for tensile architecture and has been used in the majority of the structures described in this book. Chemically both the Teflon coating and the glass fibers are totally inert. Fiberglass must be protected against moisture to avoid dangerous surface tension which would greatly diminish the strength of the tiny fibers. Since Teflon is a relatively rigid substance with minute hair cracks, moisture protection has to be provided by an undercoat of silicone. Teflon coated fiberglass is non-combustible, very easy to keep clean, highly reflective and highly translucent. Under most conditions, Teflon coated fiberglass fabrics should have a life span of 25 to 35 years. The disadvantages are its high initial cost, difficult handling due to stiffness of the fabric, its low tear strengths, and its highly non-linear behavior at low stresses.

In the 1980s Dow Corning developed a silicon-coated fiberglass fabric, which was used in a considerable number of building structures, a process in which I played an active role. Silicone (Si) is one of the most abundant elements on the crust of the earth and forms the basis for both the fiberglass threads of the fabric and the Silicone rubber of the coating. This similarity in chemical structure allows the design of highly translucent fabrics. The water protection provided by the silicon rubber coating assures a long life of the fiberglass. Cost and handling of Silicon/Fiberglass are somewhere between that of Teflon/glass and PVC/Polyester.

Past applications of this material experienced two problems. One is that the silicon rubber surface attracts dirt which cannot be removed. As a result, outside exposed fabric surfaces become unsightly. Interestingly, this has little impact on the translucency. The spaces under the fabric enclosure remain bright and clean. Therefore, a surface treatment or compatible coating has be found to deal with this problem.

The other problem relates to the fact that silicon is not a thermo-plastic material and can, consequently, not be heat sealed (welded), as most of the other fabric material can. It needs to be glued. The gluing process is time and space consuming, and it requires control of the humidity of the environment. Therefore field seaming is difficult. Shop seaming is costly. These and other aspects, unrelated to the fabric matcrial, lcd Dow Corning to abandon the fabric construction area, just as the material began to become successful.

I believe that the existing problems can be solved, and that - because of its superior translucency, long life expectancy, and good economics - Silicon/fiberglass is a prime candidate for a leading architectural fabric.

The latest material to enter the tensile structure field is polyethylene. As a vapor barrier and water proofing material it has been around for many decades. For these purposes it is available in sheets of 4 or 6 mil thickness. While the fabric materials discussed before cost between $1 and $5 per sq. ft., polyethylene sheets sell for 3 to 5 cents. Polyethylene, even in the form of thin membrane

The Poughkeepsie Mall cover is a silicone coated fiberglass roof, 18 years old at the time of this photograph, demonstrating the feasibility of this material for permanent structures.
[5.21]

An air-inflated roof formed by two layers of polyethylene sheeting, supported by steel arches, span 30'. The mem - brane cost is just pennies.
[5.22]

Materials for Tensile Structures

Experimental structure at CCNY, using NovaShield fabric, a high density woven polyethylene fabric coated with polyethylene, produced by Intertape Polymer. [5.23]

Sports facility covered with NovaShield polyethylene, a very inexpensive fabric which has many of the properties required for permanent structures. [5.24]

For smaller architectural installations pre-engineered fabric structures are available. This one is by ShadeSails, also made of polyethylene fabric. [5.25]

sheets, has considerable strength. Therefore the material has been used for many years to cover green houses for agriculture. While the sheets are not clear, they admit as much 95% light. Air supported and arch supported greenhouse structures are popular on a very large level, worldwide. The polyethylene membrane needs to be replaced after approximately three years, which is no problem at the very low cos of the material.

The new development is the production of high density polyethylene fibers which are the basis for woven and knitted fabrics. They are also coated with polyethelene. Fabric of sufficient strength for architectural tensile structures cost approximately $ 0.25. In the last few years this new material has been used for agricultural and industrial buildings of considerable size. Most recently, handsome tensile structure canopies are offered for commercial and private applications using a variation of this material. The finished products are offered at costs which are a fraction of the cost of present day tensile structures as part of permanent building projects. This indicates great opportunities for the future of architectural fabric structures.

An ideal fabric would combine all the best properties of the materials known to date. To achieve the required strength is not a problem. It is essential, however, that such a fabric is pliable to make fabrication and construction easy, and to avoid reduction in strength during the construction process, characteristics shown in polyester and polyethylene. For architectural applications it a reliable surface finish is important which always looks good and requires little maintenance. Teflon or Tetlar finishes are best at that. High translucency is desirable as in Silicon coated fiberglass. Live span is important depending on the cost of the fabric material and the relative ease of replacement of the membrane. Cost needs to be drastically reduced. The new polyethylene materials give hope that this is actually possible.

I have no doubt that in the near future fabric materials become available that will come close to all or most of these characteristics. New synthetic fibers, improved weaves (or knits), refined coatings, and, above all, higher volumes will be some of the means of achieving this. Automated cutting and seaming, simpler detailing, and easier construction methods will be other steps in the evolution of fabric as a common structural material in architecture.

*King Fahd Stadium, Riyadh, Saudi Arabia during construction. 1985.
Inside and outside are different images of one sculptural ribbon enclosing the space. For scale: detect the worker on top of the leaning strut at right. [6.01]*

6

20 Years of Fabric Structures: 1973 to 1993

In hindsight I consider myself very lucky. In a period of only twenty years, more than forty fabric structures have been built in whose evolution I was instrumental as designer and engineer. I have never thought of these structures - nor, indeed, of the many designs which were not built - as part of a larger objective. Up to the most recent (the Denver airport roof), I have undertaken each as just another job. Each has required the best that I could give, drawing on the varied capabilities I could offer. Beyond serving the specific function of each project, my objectives were always the same:to help create structures whose form would be strong, clear, intuitively

This 90 m (300 ft.) span concrete bridge at Cologne, Germany was my first major engineering project, carried out under the guidance of Dr. Bay of Wayss & Freytag A.G. Frankfurt/Main. Completed 1957. A recent visit (2004) showed it to be in excellent shape. [6.02]

University of Virginia Field House. The 87 m diameter concrete dome is made of site precast shells and arches. I designed this structure when I was an associate of Severud Associates, NY. Completed 1965. [6.03]

understandable and meaningful; spaces that people would love to be in, and buildings that would not only least hurt the natural environment, but also make use of the gifts that nature offered. No matter what my position on the team, as a prime designer or as a sub-consultant, I would invariably act as the responsible guardian over the impact of the tensile structure on the building. For me, the ultimate test was always the visual strength of the finished building: If it is good, it looks good!

In the long preparatory years as a structural engineer I had learned everything there was to learn about structures: their forms, the way they work, and how to build them. I had designed bridges, domes, shells, high rise buildings, and cable structures. I had started as a bridge

This 1961 tensile roof structure for the Unitarian Church in Hartford, Connecticut, represents a transition: a cable net supports a wood roof. The architect was Victor Lundy. [6.04]

20 Years of Fabric Structures - 1973 to 1993

and special structures designer with Wayss & Freytag, then one of Germany's most prestigious design and construction firms. I had spend two years designing reservoirs, water towers, power stations, and turbine foundations in Iran. And for eight years I had been a member of Severud Associates in New York, the structural engineering firm which had been responsible for many innovative building structures of the 1950s and 1960s. There I had participated in the design of the St.Louis Arch, the Marina City towers of Chicago, and a long list of inventive long span structures. Building structure was therefore the medium available to me in order to advance from the defined realm of the artisan to the uncharted world of the artist. Fabric structures, with their total dependence on form, gave me this opportunity.

Although I had worked on fabric structures before, my intensive involvement began with the formation of Geiger Berger Associates in 1968. David Geiger's ideas of an air-supported structure for the US Pavilion at the Osaka World's Fair of 1970 had helped his team win the design competition and make him the structural engineer of the project. He needed a firm to carry it out, and asked me to join him as his partner.

Six years my junior, David's first job out of graduate school had been on a team which I headed at Severud. After a year he resigned to go to Columbia University for a doctorate degree and a teaching career. Now he needed my capabilities and experience, and I needed his gift for creating opportunities.

With David as a driving force, the success of Osaka led to an explosion of long span air-supported roof projects. These roofs cost half as much as rigid domes, and could be built in half the time. When finished the space inside was full of daylight. Stadiums covered with air-supported fabric roofs became the fashion.

Although I spent much of my time assisting him in the design of his structures and in managing conventional projects, my main interest was the development of tensile structures. The idea of depending on mechanical equipment to hold up a structural system bothered me from the beginning: that's how we fly air planes, because we have no alternative. But in spite of all care, they crash. Building structures, on the other hand, are static.

The US Pavilion at the 1970 World's Fair in Osaka, Japan, was the project which started Geiger Berger Associates. Its success led to many air-supported and tensile structure projects. [6.05]

They stand up simply because they are shaped and proportioned to do so. To find the forms which would reliably, efficiently and elegantly make fabric structures stand up became my driving interest. But it took five long years before the first one was built.

In 1983, David Geiger and I parted our ways, and I founded Horst Berger Partners, which remained in existence until 1990. The Denver Airport project and the New York Aquarium roof were designed later, however, when I was a consultant with Severud Associates, using my software, experience, and some of my critical people. My more recent work is handled by Light Structures Design Consultants/Horst Berger, in partnership with DeNardis.

In the period covered by this chapter, few others in the design profession were seriously engaged in tensile architecture. Frei Otto, who had opened the mind of the architectural world to the fascinating possibilities and the sparkling beauty of tensile architecture, had just seen the completion of the Olympic Stadium roof in Munich. It was a masterpiece and a great landmark of architecture. But its excessive construction cost - many times that of our contemporary air-supported fabric domes and, for that matter, of conventional structural systems, - stalled tensile architecture in Europe for many years. In Japan building codes restricted fabric tensile structures to their use in temporary facilities. Therefore the many imaginative designs by Tsuboi, Taya Kogio Corporation and many others remained at the fringes of architecture.

At any rate, post-modern architecture with its anti-structural attitude was not a fertile ground for a development in which structure was the dominant formative principle. The decorated-box approach to design, in which the semi-circular arch was the most exciting geometric element, was the antithesis to the use of warped-surface sculptures whose geodesic forms were derived from the non-linear lawfulness of the new physics of "*chaos*".

Only in more recent years has there been some wider interest in tensile architecture. Architects such as Michael Hopkins and engineers such as Harold Mühlberger, Massimo Majowiecki, Mamuro Kawaguchi and Mathis Levy have produced significant fabric struc-

Canada Place in Vancouver, B.C., my last project at Geiger Berger. The architect was Eb Zeidler. [6.06]

The roof of the San Diego Convention Center by Horst Berger Partners with architects Arthur Erickson Assoiates. [6.0.07]

20 years of Fabric Structures: 1973 to 1993

The Mitchell Performing Arts Center at the Woodlands, outside Houston, Texas, was finished in April of 1990, the last Horst Berger Partners project. Appropriately, for this project I was in charge of both the architectural and the engineering design. [6.08]

ture designs. Michael Hopkins' exquisite designs, in particular, follow a philosophical approach which is quite different from my own.

Common to all of my structures is an attempt at purity of form. Some critics may read this as a sign of the narrow mindset of the "engineer". Yet I believe that clarity of purpose, simplicity of form, and directness of expression can be solid foundations for an architectural design philosophy in which the quality of its art is tested by the criterium that "*nothing needs to be added and nothing can be taken away*".

This, to me, is a critical aspect of my work. Its demonstration is a strong motivating factor for writing this book. It is the main reason why I limit this part of the book to the description of structures in which I have been personally involved.

At this point let me add a note with regard to my role in the creation of the structures illustrated in this chapter. I was directly and critically involved in the development of them all. In a number of cases, as in the Mitchell Performing Arts Center, I was the principal designer of the facility and was responsible for its architectural design. In most cases, I created the tensile structure concept. Some initial concepts came from others. The detailed structural design has always come from my office.

It is my opinion that there is too much emphasis on personal credit in the architectural profession. In reality, almost nothing is designed by one person alone. Architecture, by its very nature, is a team activity. What matters is that we use the talents we have been given, to contribute as best as we can towards creating a built environment which makes life livable, hopefully enjoyable, at times enchanting. If the structures described in this chapter make a contribution towards this end, then it was worth the effort.

Computer generated perspective of radial cable net representing the roof of the proposed Interama Amphitheatern in Miami, Florida, 1972. [6.1.01

6.1
The Way of the Spider

Of all the elements in nature water is the gentlest and the strongest. Water breaks down mountains, puts out fires, nourishes life.

Among plants, trees are the most powerful and efficient structures. The bundled tubular construction of their trunks and branches make them light and flexible, provides capillary passages through which the nourishing water, rising up from the earth, reaches all parts. Attacked by wind, trees flex, offering least resistance to the force which could otherwise harm them. Leaves become air foils, branches hug tight, shrinking their span, reducing the stress that could break them. Having dropped their leaves for the winter, the bare branches droop under the load of the snow, thwarting the built-up of destructive forces.

The finest engineer in the animal world is the spider. Her net, gentle like the water, flexible like the tree, is a marvel of construction in its simplicity and sophistication. Her method of construction is efficient and methodical. Her eight legs and six spinnerets are advanced equipment for the job. (Only the female has this capacity. Baby spiders of both sexes make beautiful toy nets, postage stamp size, but the males lose this ability when they grow up.)

Building a net starts with "knitting" a kite, and adventurously sailing off into space in search of another anchor. Once connected, a few additional passes of silk line reinforce the "bridge" from which the net construction originates. The spider now attaches a new line and sails from the middle to a third point below, thus defining the hub of her web. Now she adds more lines, making sure that the hub is stable. At the same time she builds the frame, outlining the size of her net.

Once the hub is firm, she adds the rest of the radial lines to the net, one next to the other, with industrial efficiency. She then starts the construction of the rings. She begins by laying down a spiral "scaffold", consisting of a widely spaced set of spiral lines. Reversing her direction, she then installs the closely spaced lethal rings with their sticky surface, eating up the scaffold as she moves along, and carefully avoiding to step on the sticky rings. After applying the sticky spray, she "twangs" the ring lines like a violinist playing the piccicato. The vibrations cause the line of sticky material to break up into a row of drops which will resist dirt and remain sticky.

In the process of building the net, the spider uses her multiple legs to test the tension of the lines, replacing or adjusting them when necessary. Her test method is not unlike the most common procedure used for testing high strength cables. In this test procedure, two points of the cable are held fixed and a third point, half way between them, is pushed laterally, using the magnitude of lateral deflection as a measure of the force in the cable.

Once completed, tested, and accepted the web is a structure of great beauty. It is invisible to insects, so they fly into it and are caught, their light bodies adhering to the sticky drops of the rings. The elastic net absorbs the impact of the flying insect by flexing and deforming. As the net deforms, its capacity to resist load increases while the load intensity caused by the impact of the flying insect decreases. The net, never experiencing large stresses, returns to its original flat shape, holding the insect for the spider to rush in and wrap it up into a neat nutritional package.

The spider, leaving a "telephone" line attached to the net, returns to her perch where she

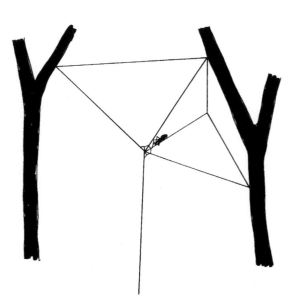

Spider Web Construction: Fixing the hub and the frame.
[6.1.03]

Spider Web Construction: With frame and radials completed, the spider installs the rings. The loose scaffold, built first, is replaced by dense, sticky rings.
[6.1.04]

Orb web of the Garden Spider.
[6.1.05]

waits for the tug that will signal the arrival of her next catch. The next day, unhappy with the tears and rips, she will remake her tent with the same imaginative perfection, retaining any useful parts of the old structure.

There is, of course, a great variation in the form of spider webs, and the techniques used by various kinds of spiders. But the geometric resemblance between the orb webs described here and radial tent structures, and the sophistication of the spider's construction process, is forever fascinating to me. Watching a Garden Spider at work is indeed a special treat.

The first of my fabric tensile structure designs to be built was, indeed, a radial tent design. And I arrived at its configuration without thinking of spiders; in fact, without any preconception as to what form of structure would be most appropriate.

It began with a phone call in the summer of 1972, asking me to come to Warner LeRoy's uptown Manhattan design studio. The studio's purpose was to design a new wild life and amusement park called "Great Adventure" in New Jersey, an hour from New York City. When I got there, Warner, a stocky ebullient man dressed in maroon pants and fire engine red trousers, offered me coffee in a tall, gilded china cup of which he had boxes full on a shelf. He was bubbling with ideas and images that seemed to come straight from World's Fairs at the turn of the century.

What he wanted me to do was to design four large stores covered with candy-striped tent roofs. Although basic, they were to become my first opportunity to realize the tensile structure ideas I had worked on for years.

Each of the four identical structures was laid out on a square plan with 21.4 m (70 ft) side lengths. The four one-story walls were topped by a single-pole tent-like roof with 24 striped fabric panels in a radial configuration. The colors were red and white, blue and white, yellow and white, and orange and white.

As usual, there was no time: we had three days to make a proposal for the design and six weeks to carry it out. The question was how to develop the exact shape and curvature of the tensile structure so it would be in equilibrium for a given pattern of stresses. Because I had no mathematical solution to generate such an equilibrium shape, I resorted to model making as a way of getting started.

I made a small cardboard model of the four walls. In the middle of each wall I attached a curved rib bowing inward the way the tent would in these locations. These would form the edges of four hyperboloid fabric panels which would cover each tent, reducing the structure to

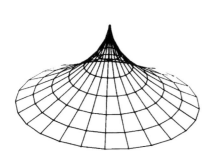

Circular Tent: all radial lines are the same; all rings are circles.
[6.1.06]

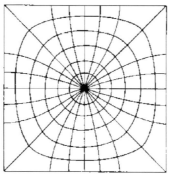

Orthogonal stress line diagram for a square stress field.
[6.1.07]

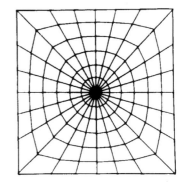

Radial Tent Diagram. Radials are equally spaced (15°). Rings are super-circles, varying from a circle to a square. [6.1.08]

The Way of the Spider

Inside, the candy striped pattern dominates. The red fabric gets very warm in the sun. The red color distorts the appearance of objects inside. Therefore, I prefer to keep fabric enclosures white. [6.1.12]

a combination of four point structures - the simplest of all tensile structure elements. Stretching a piece of panty hose over the cardboard model I found that the fabric never touched the four ribs, demonstrating that they were not required and that this roof shape represented a true tensile structure form. The ribs were not needed.

Taking Warner's candy-striped tent design as the literal basis, I arrived at a radial geometry plan layout not very different from a spider web. In the resulting net, radial lines represented radial cables. They were straight in the plan layout, spaced at equal angles of 15°. Ring lines, located on horizontal planes, represented the fabric stretched between the radial cables. A mast, pushing the hub of the net upward, would give it its three-dimensional shape, the one basic deviation from the spider web.

The challenge was to find the exact geometry of this tensile structure shape, in which the fabric and cable forces would be predictable and the surfaces would be smooth and regular. I had found that an approximate shape could be derived simply by drawing sensible force flow patterns in plan and in section. The ring shapes varied between a circle at the top and a square at the bottom. The radials were variations of the same basic form: a curve bowing inward towards the mast. It wasn't too difficult to improve the geometry graphically by re-drawing

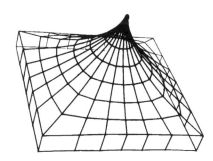

Radial tent geometry for square layout with 24 radials and 5 rings (Great Adventure). [6.1.09]

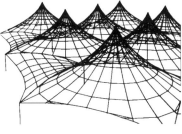

A great number of structures can be composed using the radial tent as a principal module. [6.1.10]

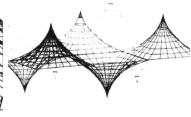

By including reversed tent shapes, even more exciting forms are possible. [6.1.11]

the lines until all the node points of the rings met all the matching node points of the radials.

This process led me to the idea for a mathematical solution. I could start with an approximate net geometry, giving each node a set of starting coordinates. The condition I could use for checking and adjusting each node point was the requirement that the internal horizontal forces needed to attach the rings to the radials in each node had to be equal and opposite in order to achieve equilibrium in the net. These internal forces are a function of the horizontal distance of the node from the mast. When the node moves outward the force increases on the ring and decreases on the radial. Therefore, equilibrium could be established at the nodes by moving them in or out along a horizontal radius line anchored to the mast centerline. What made the process easy was the discovery that the correction could be done in steps at one node at a time, assuming for that particular step that the four adjacent nodes were fixed. Making this adjustment for every node in the net, one at a time, and repeating it until no further corrections were required, led to the final shape in which all forces were in equilibrium. In mathematics, this process is called *iteration*. It is a very powerful mathematical tool, particularly suited for use with high speed computers. Its application to orthogonal net systems is described in more detail in *Chapter 8*. Iteration works if every subsequent step moves progressively closer to the final position. In tensile structure nets this condition is automatically fulfilled.

The programming of this "formfinding" process was carried out by Dr. William Spillers. By adding to it the powerful non-linear analysis method he had developed earlier provided us with the tools for finding the proper shape of this tensile structure and simulating its behavior under any load condition such as wind and snow. We could even produce the precise cutting patterns for the roof membrane and determine the exact length of the cables. We would know what stress any component of the structure would see under load. And we would know how much any point of the structure would deflect. Beyond that we now had mathematical tools which would apply to a great variety of complex structural forms.

The structure of the four buildings at "Great Adventure" consists of steel framed walls, a central tubular mast, and the tensile structure membrane. The

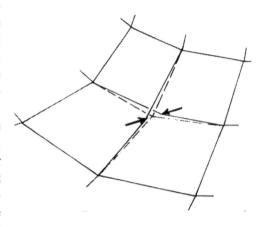

In any one node point of the net, internal forces act between a ring line and a radial line. Moving the node in the direction of these internal forces - in or out - will decrease one force and increase the other. Moving the node until the two forces cancel each other will result in the correct geometry of the node. [6.1.13]

One of the store units at "Great Adventure", a wild life and amusement park in New Jersey, 1973. [6.1.14]

upper edge of the walls is a rigid frame made of steel girders spanning horizontally. This frame forms the lower edge of the fabric roof. A six foot diameter steel ring supported by the center mast forms the upper edge. 24 cables, 5/8 in.(16 mm) in diameter, span between this ring and the steel beams of the edge frame. The PVC-coated polyester fabric fits the cable shapes with a uniform bi-axial stress of 25 lbs./in in all parts of the membrane.

The upper ring, with the fabric attached, was designed to slide up the mast. Jacks on top of the mast were used to introduce the prestress forces. The proper force was achieved with the ring elevation less than an inch from its theoretical position, an amazing level of precision for a structure of this size. It proved that our mathematics were right. And it demonstrated the outstanding fabrication quality by Birdair Structures, the structure's fabricator and erector.

Proud of this achievement, I called "Engineering News Record" and suggested that they visit the site and do a story. They visited the site; they didn't do a story. "These are just tents", they told me.

I knew that despite their striped, county fair like appearance, these simple structures represented a new approach to architectural form and technology in which form literally follows function: the shape of the structure is not arbitrary but derives inevitably from the conditions of internal equilibrium of a membrane stressed between given supports. Shape is an essential aspect of structural strength. Art and engineering are inseparable. These were not "just tents", although their colorful stripes and simple forms seemed to say so. They were tensile structures. The proof was that they had the right stress when they had the right geometry, and the fabric had no wrinkles.

What neither ENR nor I could know at the time was that only a few years later a direct derivative of the "Great Adventure" radial tent form would become the primary building element of the world's largest roof structure, covering the 105 acre Haj Terminal of the Jeddah International Airport. Today, as I work on this book for its second edition, the Haj Terminal is about 25 years old and in good shape.

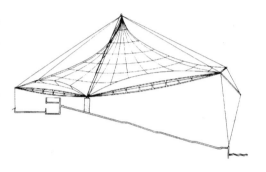

Cross Section of the Interama amphitheater roof.
[6.1.15]

The Way of the Spider

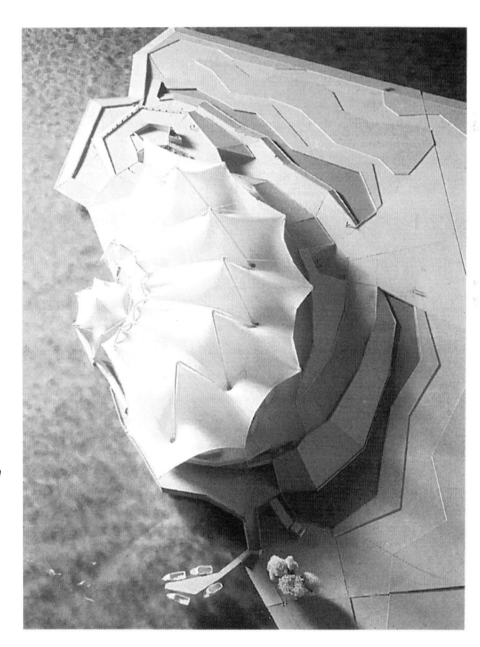

Model photograph of the Interama design [6.1.16]

The Folklife Pavilion at the 1976 bicentennial celebration in Philadelphia, one of numerous fabric structures erected for the event. [6.2.01]

6.2
Celebration in the City

Preparations for the bicentennial celebration of 1976 revealed a great deal of interest in tensile structures. They were quick to erect, inexpensive, and they had a festive flair. Most importantly, they were temporary. No permanent commitment to a new architectural form had to be made. In architecture, as my wife observes, new things start with "beer and circuses".

There were scores of inquiries. We produced design concepts for all of them. But only one of the many projects became reality. It was called Phila 76, and was to serve the celebrations planned for the city of Philadelphia. Opening was scheduled for July 4th, 1976. The planned structures included three major pavilions of approximately 2,300 sq.m (25,000 sq.ft.) each and a considerable number of smaller, mostly modular structures spread over the downtown area, housing mini-stages and food pavilions.

Two of the three large pavilions were carried out. Barry Eisworth of the architectural firm of H2L2 drew elegant concept sketches to illustrate their ideas. These were the guides for the conceptual designs which I generated with the help of stretch fabric study models. They became the basis for the final design of the structures which evolved step by step in the team design process.

The most prominent of these structures was the Independence Mall Pavilion. It sat in a highly formal and important city space, facing Independence Hall in which the Declaration of Independence had been formulated.

Light Structures
Structures of Light

Independence Mall Pavilion at the 1976 bicentennial celebration in Philadelphia.
[6.2.02]

Concept sketch for the Independence Mall Pavilion by Barry Eisworth of architects H2L2. [6.2.03]

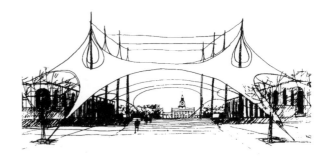

Celebration in the City

No ring is needed to terminate the fabric at the peak. The fabric wants to slide up, not down. The varying slope of the radial cables is handled by arranging connector plates like a pipe organ. [6.2.05]

The roof's support geometry was carefully coordinated and integrated with the paving pattern and the arcade design of Independence Mall (which is located on top of a three-story parking garage), so that only one stone slab had to be removed at any one anchor location in such a way that it could later on be reinstalled.

Supported by eight 18.3 m (60 ft.) high masts, it covered and area sufficient in size to protect a stage and 1500 seats against rain. The 300 mm (12 in.) diameter pipe masts were sloped outward to improve the structural balance and express the flow of the forces. Sweeping edge catenaries, spanning the full width of the structure, arched over the front and back openings. Radial cables, curving three-dimensionally, reached from edge to mast, from mast to mast, and again from mast to opposite edge. The connector fins at the mast-top connection were staggered like organ pipes to follow the different slope of the radials, whose common theoretical origin was a point in the center of the mast. A plate on the top of the mast provided the seat for the jacks which were used to stress the membrane. No ring was required at the top edge of the fabric. Counter to expectation the roof membrane wants to slide up rather then down.

A PVC-coated polyester fabric with 10% translucency provided just the right light level under the roof in the daytime, while reflecting most of the heat. This material was also economical and easy to handle during fabrication and construction.

The Folk Life Pavilion, located just across from the

My study model for the Independence Mall Pavilion. [6.2.04]

This children's climber was one of many different structures we designed for Philadelphia's bicentennial celebration. We shaped it from a standard polypropylene net. [6.2.06]

Municipal Art Museum on Eakins Oval, was a very different structure, despite the fact that it was also supported by two rows of masts. It had eight masts located along each longitudinal edge. Ridge cables, spanning across the structure between the mast tops, transmitted the downward loads to the masts and required diagonal tie-back cables to resist the horizontal pull. The valley cables were anchored downward midway between any two masts along the same peripheral support lines.

Continuous concrete grade beams below the ground surface made this arrangement simple and economical. The resulting shapes resembled the pointed arches of a gothic cathedral, an impression strengthened by the flying arch geometry of the diagonal anchor cables.

The other important distinction was that the fabric membrane shape was based on completely different principles. This was not a radial tent system. There were no radial cables. In this structure, cables were limited to ridges, valleys and edges. The hyperboloid elements, generated by filling the spaces between these cables, had enough curvature to resist both upward and downward loads with fabric stresses below the safe limits. This proved to be an efficient and practical arrangement which I adopted in the design of most of the later projects with regular mast supports.

A third major pavilion, covering an amphitheater which would use the monumental exterior stairs in front of the Municipal Arts Museum, was designed and a bid submitted. In order to keep the project within the Phila 76 budget, it was not built.

Numerous smaller structures covering a children's pavilion, mini-stages and food pavilions adapted an A-frame supported structural system for which I held a patent. These are described in Chapter 6.5.

The fabric structures designed and built for the 1976 celebration were a big success for the city of Philadelphia. They had won acceptance and applause for their gentle sculptural forms, the floating lightness of their translucent membranes, their sensitive response to the existing elements of the city environment, and the visual clarity of the structural force flow, all effortlessly achieved by the building's formative elements without the need of obstructive structural armament. They enriched the city's spaces, adding a note of cheer, taking nothing away from what existed, and claiming no pompous importance of their own. No wonder the city wanted to keep them!

Unfortunately, the structures had been designed for summer conditions only in order to cut cost. They were not safe for snow loads, I told the city and its lawyers.

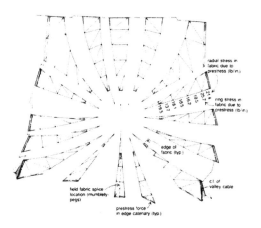

Fabric patterning elements for half of an exterior bay of the Independence Mall Pavilion. [6.2.07]

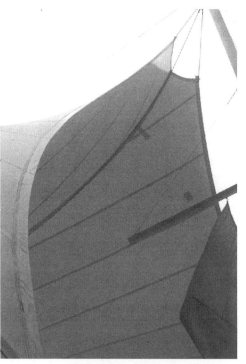

The edge catenaries of the Folk Life Pavilion had this beautiful cork screw shape which was unexpected. [6.2.08]

Celebration in the City

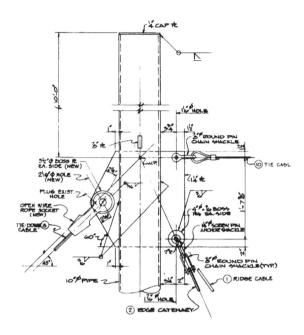

Connection detail of the mast top of the Folk Life Pavilion.
[6.2.09]

The Folk Life Pavilion, reconstructed and reinforced for winter loads, erected at Penn's Landing in 1978.
[6.2.10]

But none of my warnings had an effect. The structures were kept up for one winter, for another summer, and for most of a second winter. It was the day after the failure of a large arena in Hartford, Connecticut, that we were asked to write specifications for the removal of the structures. Tomorrow, please!!

A year later we were asked to find ways of re-enforcing the Folk Life Pavilion to be erected in a new location on Penn's landing. It remained there quite a number of years, blending in with the masts and rigging of old schooners moored alongside.

The impact of the Philadelphia bicentennial tensile structures on the future of tensile architecture was critical directly and indirectly. A young architect who was on the design team for one of the un-built Philadelphia tensile structures joined the New York office of SOM soon afterwards. It was he who first suggested using a tensile structure solution for the Jeddah Airport project. The Folk Life Pavilion, on the other hand, became the model for the design of Canada Harbour Place in Vancouver.

Though it was built nine years later in another city, another country, at the opposite end of the continent, Canada Place in Vancouver, British Columbia, has much in common with the Philadelphia fabric pavilions. Therefore this is a good place to talk about it.

By the time of the World's Fair of 1985, which was held in Vancouver, fabric structures had taken long strides towards acceptance as architecture. The Jeddah Airport had been completed. Half a dozen stadiums had been covered with air supported structures, the most recent one, B.C. Place, in Vancouver itself. Bullock's had put fabric roofs over two of its new department stores in northern California. One of the two had been selected as one of Time Magazine's 10 best designs of 1980.

Thus a fabric roof was selected for the new convention center facility of the city of Vancouver, which was a permanent building. That it began its life as a pavilion of the 1985 World's Fair under the name of Canada Harbour Place made this courageous selection politically easier.

The sail-like character of the Folklife Pavilion was an appealing concept to base the design on. And architect Eb Zeidler of Zeidler/Roberts in Toronto liked the idea. But he wanted to express the orientation of the street system by running some cables at the skew of 45%. I suggested that we orient the tent units themselves that way and offered to try it out in a study model.

The skewed plan geometry of this end supported roof system made it the hardest model I ever built. I had only a few days to get it done since our next meeting was scheduled a week after the first. I returned to Toronto with the model in my tennis bag, fearful of problems with

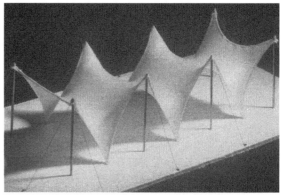

The study model for the Folk Life Pavilion reveals my love of Gothic forms.
[6.2.11]

The study model for Canada Place shows the skewed orientation of the fabric bays and the intriguing surface shapes which result. Even the model was not easy to built.
[6.2.12]

Celebration in the City   91

Canada Place with its sails, masts, and cable stays, blends naturally into the marina images. [6.2.13]

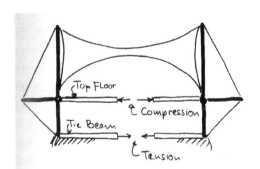

This diagram of the Canada Place structural system shows how the horizontal forces, caused by the ridge and valley cables, are transmitted down to and resisted by the floor system of the building below. [6.2.14]

Canadian customs because I still had no work permit. The bag also contained scissors, stretch fabric, pins, wire, pliers, and other paraphernalia needed to work on the model in Eb Zeidler's office to incorporate any ideas the meeting would produce. Fortunately I did not have to explain to customs what I intended to do with all of this as a common tourist.

The building sits on an old pier projecting out into the water. It is exquisitely designed and detailed by Zeidler's office taking its visual clues from the ocean liners which moor along side it. The fabric roof on top of the building resembles the rigging of a giant sail ship. It spans 55 m (180 ft) across the main exhibition hall. Measured diagonally along the ridge the span from mast to mast is 73 m (240 ft.) The five skewed tent modules terminate in the circular structure of the I-Max Theater at the waterside end. The bulk of a hotel tower forms the visual anchor on the land side.

Being part of a permanent building, the roof has a double skin of Teflon-coated fiberglass. The outer fabric is the structural membrane and the waterproof enclosure of the building. The inner liner, specifically designed by Chemfab of Vermont for the combined function of thermal insulator and sound absorbing ceiling, hangs just a foot below the outer fabric. The air space between them is a critical element in the performance of those functions. The two skins also help reduce transmission of external noise. When you see the sea planes take of and hear the roar of their engines you understand why this is important.

Anchoring the external masts safely and economically is an important aspect of this design. We

Light Structures
Structures of Light

The aerial view shows the Exhibition Hall roof and its extension over the I-Max theater in the foreground.
[6.2.16]

Celebration in the City

For safety, pairs of cables hold the tops of the tapered steel masts in position.
[6.2.15]

found an efficient solution to this structural problem by engaging the floor structure of the building below by using rigging methods not unlike those which stabilize the masts on sail boats. Stay cables attached to the tops of the main masts are held outward by horizontal outriggers. Below these the cables slope inward returning to an anchor point vertically under the centerlines of the masts.

This arrangement causes a couple of balanced horizontal forces. The upper one puts the concrete floor structure of the convention floor in compression. The lower one causes tension forces which are resisted by post-tensioned concrete grade beams which are imbedded in the earth of old piers. As a result of this arrangement, only the vertical loads from the roof reach the foundation caissons. The light weight of the roof structure keeps these vertical foundation loads much smaller than those caused by any other roof system.

A pair of cables is used for each tie-back to increase the safety and make it easy to replace them.

The fabric panels closing off the ends of the modules form huge spinnakers which are attached directly to the edge catenaries of the roof, reinforcing the image of a huge luminous sail craft which has just docked, bringing a joyful message to the beautiful and lively city of Vancouver.

Under one of the ten modules of the Haj Terminal at Jedda International Airport, during a visit in 1981. The huge tent structures act like trees in a forest, moderating the impact of the desert sun. [6.3.01]

6.3
A Forest in the Desert

In the spring of 1977, the New York office of architects Skidmore Owens & Merrill was asked to design a terminal structure for Jeddah airport, in Saudi Arabia, to serve the enormous numbers of pilgrims who arrive by air every year during the the Holy Month of the Haj on their way to Mecca, just 45 miles away. With up to one hundred jumbo-jets arriving on anyone day during the Haj, 24 hours were estimated for handling and processing these pilgrims, and transferring them from aircraft to buses. To protect them against the heat of the Saudi Arabian desert required the largest roof structure ever built; indeed, a roof with special environmental properties.

This roof, covering a space as large as 13 Astrodome-size stadium buildings, became a turning point for fabric tensile structures; it was also to become the most challenging project of my career.

My involvement began with a telephone call from Roy Allen, then the senior design partner in SOM's New York office, who was in charge of the project in its conceptual stage. I had worked with him before as structural consultant on several major projects (one of them was the Annenberg tower at Mount Sinai Hospital in New York City). Roy asked me to join his team in order to investigate the potential of a fabric structure solution.

Rendering of the Jeddah Haj Terminal by Owens Corning. The 105 acre roof consists of 10 modules, each of 21 tent units. [6.3.02]

The year-old bicentennial fabric structures of Philadelphia (described in Section 6.2) had evidently left an impression.

The structure's main purpose was to provide comfort and protection against the heat of th desert sun. A conventional, enclosed building with insulated walls and roofs, doors and windows, air-conditioning and ventilation was out of the question, due to the large area. An open roof could serve the purpose, demanding a fraction of the time and money, and requiring minimal maintenance during the 11 months of the year when not in use. The roof's main purpose would be to reduce the impact of the sun; in this respect, a fabric structure had superior properties, as had been particularly emphasized by our early studies of various roof configurations.

On a typical Jeddah day, with an ambient air temperature of 35°C (95°F), a concrete or metal roof exposed to the desert sun would build up temperatures in the order of 65°C (150°F). Because of its mass, such a roof would collect and radiate a considerable amount of heat into the building, even if highly insulated. Electric lighting - necessary under an opaque roof - would add to this heat load.

By reflecting most of the sun's heat, a tensile structure using Teflon-coated fiberglass would maintain its surface temperature within a few degrees of the ambient air. As a result, the fabric's temperature would not rise much above 40°C (104°F), while its translucency would make artificial light unnecessary during the daytime. Most of the devastating impact of the desert sun would thus be eliminated, reducing the heat to the level encountered in the shade of trees. At the same time, mechanical energy would be drastically reduced.

There were other advantages in favor of a fabric solution: most components, except for the concrete foundations, could be manufactured abroad, reducing the time and cost of construction by large factors. The roof elements, weighing as little as 10 kg/sq.m (2 lbs./sq.ft) of roof surface and consisting predominantly of flexible materials fabric and cables - took up little space, and accordingly required little shipping capacity. In fact, not only would costs be lower and construction time shorter than with more conventional designs, but some very exciting design solutions would also be possible. The conceptual study resulted in the decision to use fabric structure technology for the proposed project.

By the Spring of 1977, Geiger Berger Associates had completed a number of air-supported structures, including the Pontiac Silverdome, covering a football field and seat-

A Forest in the Desert

ing for 78,000. The firm had also produced several designs for tensile fabric structures, but only the tiny Great Adventure tents (Section 6.1) and the temporary structures for Philadelphia's bicentennial celebration had actually been built. The Jeddah roof was to be 200 times the area of the larger of these, and although very little of the experience gained with air-supported structures had relevance for this tensile structure project, at the time no one else had even a fraction of our experience and design capability.

Geiger Berger had developed the necessary design tools, including powerful computer programs, and in the realization of the above structures, had demonstrated their reliability.

The Saudi Government put things in motion by deciding to go ahead with the project and retain SOM as architects for its final design. Gordon Bunshaft, recently retired from his position as the dominant partner with this giant architectural firm, decided to return and head the design. He called me to confirm that we would be his structural consultants. I was ecstatic - this was the reward for many years of patient preparation.

Four weeks later, not having heard from him again, I called back. "Sorry", he said, "we changed our minds. We'll do the engineering in-house". I was perplexed. The rumor was that the project had been moved to SOM's Chicago office, which had a strong engineering section. What had happened? How could they possibly handle the design?

I heard nothing more until early next Spring: "I hate to mess up your weekend", Owens Corning's Verne Oase said when he called from Essen, Germany. He eXplained how they had received the bid documents from Hochtief A.G., the general contractor, and to everyone's surprise, these documents - for the construction of the new Jeddah Haj Terminal - made the roof contractor responsible for overall design and engineering of the project. Two of the four competing roof contractors withdrew immediately: they did not have the design capability; and nor could they find it.

It was agreed that the Owens Corning team would stop in New York on the way back from Germany. There were just six weeks to convert SOM's concept into a feasible design and a bid there would be no weekends for a while.

SOM's drawings showed only the outlines of the building; there were no details. Nevertheless, the concept was exciting and attractive. It looked simple enough: ten identical modules were ar ranged along a central spine,

The architect's drawing show the seven units which make up the length of each module, three make up the width. To resist the lateral forces, the end modules come in pairs which form rigid frames.[6.3.03]

five on each side. The spine contained the access road for vehicular traffic, including the buses which would carry the pilgrims to Mecca. The gates for the aircraft were located at the outer ends of the modules. Each of the ten modules consisted of 21 square, radial tent units with a side length of 45 m (150 ft.). Each module was three units wide and seven units long, resulting in overall dimensions of 135 m (450 ft.) by 315 m (1050 ft.), with a plan area of 43,000 sq.m (460,000 sq. ft.). In other words, each module was a third larger than the Astrodome in Houston, Texas - a scale which was difficult to envision.

The square, radial tent units were enlarged versions of the Great Adventure shapes, using 36 radial cables to support the Teflon-coated fiberglass membrane. What was unique was their support system - the idea of SOM partner Fazlur Khan, who had taken charge of the design when it moved to Chicago. He was the structural engineer known for the design of Chicago's highest buildings: the Sears Tower and the Hancock Building. In 1980, at one of a number of pleasant breakfasts we shared during a conference in Stuttgart, Khan and I

This computer drawing illustrates the structural system in simplified form. Nets of various densities were used for different parts of the analysis. Though we used a Cray computer, the system was sometimes too large. Today you can analyze a much denser net on a laptop computer. [6.3.06]

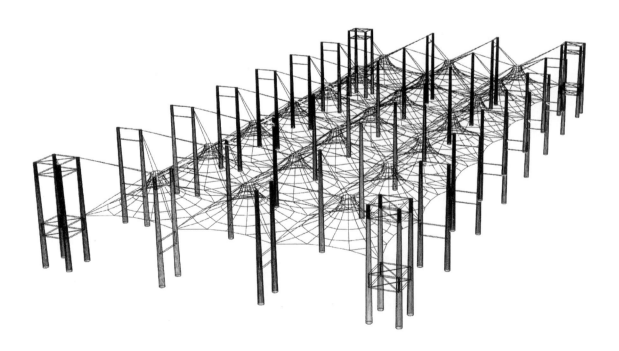

A Forest in the Desert

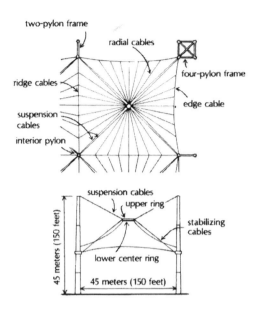

A typical tent unit has 32 radial cables. Its edges are catenary cables or valley cables. Plan view. [6.3.04]

Four suspension cables transmit the load to the pylons located in the corners. Elevation. [6.3.05]

pieced together what had happened during the design period.

Bunshaft had shown Khan his design concept, consisting, he said, of 120 m (400 ft.) spans of draped fabric shapes. Khan had told him that this would not work. Bunshaft's response was: "You are an engineer, make it work!" Khan then tried to explain that the form of a tensile structure cannot be arbitrary; that a tensile structure won't work unless you shape it correctly. Bunshaft didn't like that; in fact he was furious. "Then you design it!", he yelled at Khan. "So I did", he explained to me. That's why the design had moved to Chicago. And that's why I had never heard of it again until the time of the bid. In order to avoid central masts under the peaks of the tents, Khan placed pylons at the corners of the tent units, and made them high enough so that the tent peaks could be held by four sets of cables suspended from the mast tops. This was not a very economical arrangement, especially since the edges needed heavily reinforced double pylons to resist the large anchor forces high above the ground. Nevertheless, this design gave the space its open powerful organization, and the building its dramatic form.

We managed to complete the bid design on time: Owens Corning quoted $165 million. Months of presentation, negotiation and speculation followed. By October of 1978, the project was under contract. Now, the results of our quick bid design had to be rigorously analyzed, confirmed, finalized, and supplemented - again under the constant pressure of deadlines: there were four days to establish foundation loads; and 21 days to determine the loads for the steel pylons, which were to be fabricated by NKK of Japan, and were critical to the construction schedule. In fact, NKK had threatened to abort their $ 65 million share of the project if we missed that deadline.

We almost did! Our last computer task, the simulation of the effect of a major rip in the fabric of one of the units, had not worked out. Something had interfered within our program, and Steve Goodson, my very capable associate, decided he had to rewrite it. He was going to spend the night doing it. The next morning I couldn't reach him by phone anywhere, and when I got to his house in Greenwich Village, I found him asleep. He had worked all night and, after completing the rewrite, had fallen asleep! We made it at the last minute.

In the following months, thousands of pages of computer analysis, and stacks of engineering and patterning drawings, were produced. The work included the design of a prototype structure, which was erected in Owens Corning's plant site in Granville, Ohio. Before construction in Jeddah could be given the go-ahead, the specification required that everything should be tested on a two-unit structure to confirm the results of our computer analysis on the one hand, and overall structural design on the other. Trips to Granville became routine. The prototype construction sometimes got ahead of the design, requiring hasty adjustments over weekends, or the finding of materials that could be delivered overnight and welders who needed the extra income.

Computer work remained critical, and here we were particularly lucky: United Computer Services, whose computers we were using in time sharing at that time, had just installed a Cray, the world's most powerful mainframe computer. We were the first to get on it. My two associates, Steve Goodson and Andy Stefaniuk, who were highly capable and thoroughly familiar with our programs, told me that we would have no problems as long as the system was not too large for the machine. They assured me it was not. But they were wrong! Much of my daytime and some of my night-time during those early weeks was spent finding ways to make the large matrix system fit on the machine. Once it did, the answers came fast.

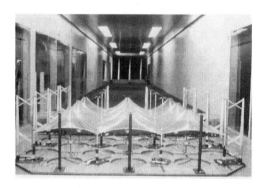

Dynamic wind tunnel testing at the University of Ontario under the direction of Alan Davenport. The tests used nine units with meticulously scaled properties. These are the only dynamic tests on fabric structures I am aware of. [6.3.07]

The prototype testing was invaluable. The structural components of the prototype units contained stress sensors connected to a central computer, so that we were able to record the stresses throughout the system, and compare them with our computer analysis. The results were encouraging: deviations were in the range of 3%, and never exceeded 5 %. We knew that the theories we were using made sense, and that the programs we had written were correct.

The size of the project also provided opportunities to get answers to some questions which had been bothering

A Forest in the Desert

Two full scale prototype units were used to test the system prior to fabrication and construction. [6.3.08]

Stress sensors connected to a computer recorded stresses in every part of the system for comparison with the results of the computer analysis. Owens Corning's Verne Oase indicates his whole-hearted approval. [6.3.09]

us for some time. Dynamic wind tunnel testing was one of them. Could a stream of air flowing over a structure of this kind cause vibrations which could build up to destructive force levels? Could the structural behavior mechanisms which brought down the Tacoma Bridge be relevant for lightweight tensile roof structures? and, in particular, were the edges of fabric roofs, held in position with flexible catenary cables, susceptible to dynamic excitations?

I did not believe that dynamic model testing could be done in a meaningful way. There were so many properties which had to be scaled down to a model small enough to fit into the eight foot section of the wind tunnel, and the scale requirements of the different properties were bound to be in conflict. But I had underestimated the ingenuity of Alan Davenport and his associates at the University of Western Ontario. Their test process was amazing, and watching the tent units deflect under the pressure of the wind - steadily, reliably, the front row moving clearly more than the second, the second more than the third - was very exciting and gratifying. The observed deflections confirmed the results of our computations. Most importantly, there were no vibrations. The absence of weight, the mixture of materials with widely different properties, the non-linear behavior of the structure, and the large surfaces being dragged through the air, all combined to dampen any dynamic build-up. Sailors had known that for ages; indeed, we had believed it, but here was the experimental confirmation.

Light Structures
Structures of Light

This 1981 construction photograph gives an idea of the expanse of the construction site. In module F, the units are being stressed. In Module G, 14 of the 21 units have been installed near the ground. Erection of the pylons of module H has started. Less than one quarter of the completed total roof structUre is visible in this photograph, taken from a ring of the completed module E, across the entrance road.
(Imagine the parts of the photo eliminated by the margins). [6.3.10]

A Forest in the Desert

Watching the real structure being built, and being part of that process, was one of the great experiences of my life. The scale of the construction site was simply astonishing. The contractor's compound was a small town right next to the site. We needed cars to get around, and we had to watch out for heavy equipment which might cross our way, or a trench that had been dug the day before. The structure has an amazingly small variety of components. The tent units consist of fabric, radial cables, edge and ridge cables. They have support rings at the top and connection elements at the corners. The rings are held by suspension and stabilizing cables which carry the loads to the pylons and their footings.

The tent units come in three variations: interior, edge, and corner units. All have 32 facets, with radial cables located in fabric pockets along the seams. Edge catenary cables define the outer edges of the corner and edge units. Ridge cables are located along their interior edges. The corners are connected to the pylons 20 m (65 ft) above ground. At the upper end of the tent units, fabric and radial cables terminate in a 15 ft. diameter steel ring - the lower part of a matched pair of rings located 34 m (110ft.) above ground. In accordance with the geometry defined by these elements and the prescribed uniform prestress in the fabric, our computer program would find the shape of the units, the lengths of all cables, and the patterning required by Birdair and Owens Corning for fabrication of each fab-

Walter Bird uses a paper model to demonstrate his method of folding the very large tent unit so they fit in one box. [6.3.11]

In Jeddah, one of the 4ft.x8ft.x16ft. tent crates is placed under the center ring. [6.3.12]

After attaching the fabric to the lower ring, the ring is lifted and the fabric unfolds. [6.3.13]

The principles of the construction process are illustrated in the perspective sketch of a corner unit. [6.3.14]

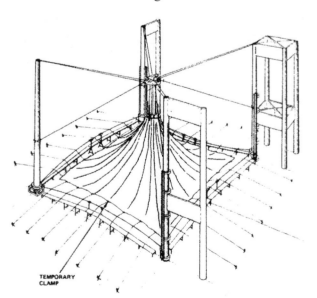

A Forest in the Desert

The modules are constructed in an assembly line process based on the design of the double ring system, located at the top of the units. [6.3.15]

*The photos below show:
The upper ring, suspended from pylons, with winches and electric jacks. [6.3.16];
Units are assembled near the ground. 21 units are raised by winches. [6.3.17];
Threaded jacking rods are used to stress the structure. [6.3.18]*

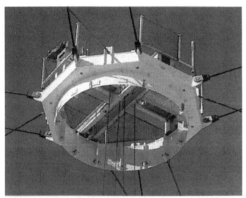

ric segment.

Based on biaxial tests carried out on samples taken from each roll of coated fabric, Birdair would compute the "compensation". (Compensation assists in the creation of fabric panels that are shorter by the exact amount that they stretch under stress during the construction process.) Walter Bird, the founder of Birdair Structures and at that time still the firm's head, had developed the procedures. Without his know-how, experience, and uncanny inventiveness, fabric technology would not have been ready for a project like this.

The double ring arrangement, an innovation by Walter Bird and Verne Oase, was the central idea which made the sophisticated erection process of the modules possible. The 21 unit module could now be assembled near the ground, where workers had easy access, and all twenty one units could then be raised and stressed together.

The rings are held by suspension cables which reach up to the 45 m (150 ft.) high pylons. Stabilizing cables stretch from the rings down to the corners of the tents, holding them steady before the tents are attached, and provide redundancy for the system under load. The cables are high strength galvanized bridge strand of the type used for decades in suspension bridge hanger cables. For further corrosion protection, the large cables are enclosed in a polyurethane sheathing

The pylons are made of tubular steel sections which taper from a diameter of 3.66 m (12 ft.) at ground level to a diameter of 1.83 m (6 ft.) diameter at their peak. They collect all the vertical and lateral forces exerted by the cables - suspension cables at the top connection of the tent units, and edge catenaries, ridge cables, and stabilizing cables at the bottom connection - and bring these forces down into the foundations.

For interior pylons, this is a simple task because all connecting cables reach the pylons in balanced pairs. Any unbalanced loading from opposing cables cause the top of the single pylon to move until the loads balance out, but the small amount of bending is an insignificant contribution to the stresses in these pylons.

The exterior pylons, on the other hand, have to resist the entire unbalanced load in one or - in the case of the corners - two directions. Under strong horizontal bending loads, double pylons quadruple in the corners - act as rigid frames, and because of the resultant heavy bending stresses, they require thicker wall sections.

This pylon configuration was a purely architectural choice. Structurally, a much more efficient solution would have been to use exterior diagonal cables, as in the Folk Life Pavilion in Philadelphia. This would have drastically reduced the cost of the pylon system. A compromise was to add diagonal cables into this pylon design, thereby retaining the visual configuration while avoiding the costly frame action.

The construction schedule was based on the erection of one module at a time. The double pylons, located between modules, had to resist the one-sided load until the adjacent module had been installed and stressed. The addition of each module changed the loads on the pylons in existing portions of the system, causing them to move at their tops. This confronted us with complex engineering problems: we had to predict the magnitude of these movements, and find ways of positioning and surveying the pylon tops, so that after installation of the last module, all the pylon tops in the system of five modules would be in their proper position.

The pylons and rings were fabricated by NKK in Osaka, Japan, and shipped on special barges to the port of Jeddah. Specially designed vehicles transported the 45 m (150 ft.) pylons to the construction site for installation. The cables and their connections were made in France, while the fabric units were produced in the United States by Birdair Structures and Owens Corning Fiberglass. They arrived, carefully folded, in crates 4 x 8 x 16 ft. With the pylons of one module erected, and the upper rings installed and stressed with the stabilizing cables, the fabric crates were placed on the ground,

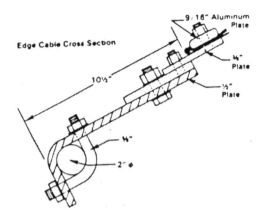

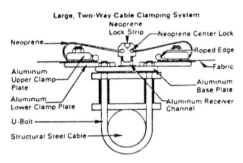

Top: Section through the clamping at the edge catenaries. [6.3.19]
Below: Section through the clamping at the valley cables. [6.3.20]

The lower connection detail of the tents to the pylons. The frame was an erection feature, allowing the units to ride up during winching. [6.3.21]

A Forest in the Desert

The construction site at night.
[6.3.22]

centered under each ring. The crates were dismantled. The top of the tent unit was attached to the bottom ring, which was in turn attached to the cables of a winch on the top ring, and the fabric unit was then lifted out, unfurled, and stretched out towards the edges. After installing the radial cables from below, the 21 units were clamped to the edge and ridge cables. The entire 21-unit module was then simultaneously raised by the winches, controlled from an electronic command post on the ground. Four stressing rods were then inserted into holes in the lower rings, and connected to electric jacks. Using this equipment, the roof system was pulled up for the last four feet and stressed. At the design prestress level, the bottom rings nestled inside the top rings. When the cables and fabric of all 21 tensile structure units were at computed stress levels, eight connector rods were installed through matching holes in the upper and lower rings.

I vividly remember docking the rings of module A (the first module). It was in January of 1980. And there were two things that worried me: Would the rings all slide into each other without trouble? and: Would the holes in the matching ring pairs all line up for the connector rods to fit? In both cases, I was relying on theo-

The rope ladders were kept short to keep everyone in shape. You had to take a running start to make it to the bottom rung. Sometimes you would miss, sliding back down the steep, slippery Teflon slope.
[6.3.23; 6.3.24.]

Under module A during construction. [6.3.25]

ry only. Apart from the forces in the fabric and cable components, and the accuracy of their fabrication and installation, there was nothing to guide the geometry of the system,

All the rings docked easily - exactly as predicted by the computer analysis. And by the time the system had achieved 80% of the full design stress level, there were only a few inches to move upward into final position and I could look up and see daylight through all of the small holes 100 ft above me. In fact, all the rods went in without forcing. Nothing is more practical and more powerful than a theory which works!

After the erection of Module A, the project moved on without a hitch, and was completed in less than three years. It contains more Teflon coated fiberglass fabric than all earlier structures added together. Although it moved tensile structure technology a giant step forward, it did not have the decisive impact on the construction market which many of us had hoped for. It did change the attitude of the Saudi Arabian government, however, which had accepted the tent image of the Haj Terminal

A Forest in the Desert

On top of module A during construction. For scale, note the two workers walking on the roof. The total roof is ten time as large. [6.3.26]

with some apprehension. For their next large roof project, the Riyadh Stadium, they specifically wanted tent shapes, which were now regarded an appropriate symbol of indigenous construction technology.

Almost 25 years later, the giant roof structure is in excellent condition, demonstrating the reliability and durability of fabric structure technology on the largest scale imaginable. It is still the world's largest roof structure, and will probably remain so for a very long time. It is an outstanding sculptural expression of structural form, but, most importantly, it comes close to nature in its gentle way of taming the heat of the sun: a veritable forest in the desert.

Light Structures
Structures of Light

Every year during the "Haj", tens of thousands of people stream through the terminal. Buses are lined up to take people to Mecca, 45 miles away. [6.3.27]

The roof, not unlike a forest, protect the Hajis from the impact of the desert heat. 6.3.28]

*Facing page :
Above: Up to a hundred jumbo-jets unload pilgrims on each day of the holy month.
[6.3.29]*

*A plane takes off over the sea of planes.
[6.3.30]*

A Forest in the Desert

112

"Florida Festival" of Seaworld, Orlando, at the time of its opening in January of 1980, glows in the night. The large radial tent unit, with its mast leaning outward, is balanced by three small tent units. At their intersection, a reversed tent form holds the structure down. [6.4.01]

6.4
Big Tops to Stay

The Jeddah project was not the only offspring of the Philadelphia tents. Among the others were roofs for two *Bullocks* department stores in northern California, and two projects for *Seaworld*, the wildlife and entertainment park company. It was the courage of the business executives in these organizations which moved the new technology forward.

The larger of the Seaworld projects, Florida Festival, was an addition to the Seaworld park in Orlando, Florida. Its purpose was to house "things Floridian": restaurants serving Florida dishes, shops selling Florida products. Architect Robert Lamb Hart had proposed a

Palm trees grew happily inside the space of the Florida Festival structure. The diffuse light, though only 16% of full daylight, provided sufficient. A trussed steel mast forms the support of the large tent unit. [6.4.02]

conventional structure, using flat roofs with metal framing. After seeing the bicentennial structures in Philadelphia, however, George Becker, who was the manager of the Florida park at the time and later became the Seaworld's president, asked me to come down to Orlando and discuss a translucent fabric structure as the main enclosure element of the planned 5,000 sq.m (54,000 sq. ft) facility.

A fabric roof using Teflon coated fiberglass produced by Chemfab of Vermont could transmit as much as 16% of the full daylight. A study by a horticulturist indicated that a large variety of Florida trees and plants would do well at such a light level - even palm trees would flourish. George Becker decided to take the risk and go ahead. We were asked to join the architectural team as design and engineering consultants and develop a tensile structure roof design.

My first idea was to combine regular tent-shaped modules with inverted ones. Such a combination would give the space the rich variation of volumes it needed. I asked the architects to change the two intersecting rectangles of the original floor plan into squares. The resulting regularity of a modular square grid opened the opportunity for a combination of tent forms, giving it the discipline needed to contain the exuberance of their dynamic shapes.

To use only two large square elements only, one pointing up and the other down, would not work because the support point of the inverted element would end up way below grade. Adding three small tent units to the large unit to surround the inverted element proved to be the answer. This evolved from my working with stretch fabric study models.

The study model also showed that the three small masts had to be vertical, while the tall mast had to lean outward, in order to effectively resist the forces pulling its top inward. A concrete edge beam, placed on an earth berm, formed the periphery, a feature which had worked well in the Osaka pavilion.

My first sight of the completed structure was exhilarating. I arrived late at night, the day before the opening ceremony. All the lights were on inside the pavilion, and it was glowing in the Florida night sky. I climbed on

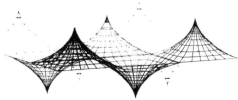

Computer generated radial tent design using a combination of upright and inverted tents. [6.4.03]

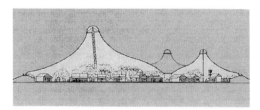

The section through the building shows the composition of tent units and their dynamic shaping of the space. [6.4.04]

From the air, the simple layout of the roof units is clear. Canopies formed by sweeping edge catenaries form the entrances to the building. [6.4.05]

Big Tops to Stay

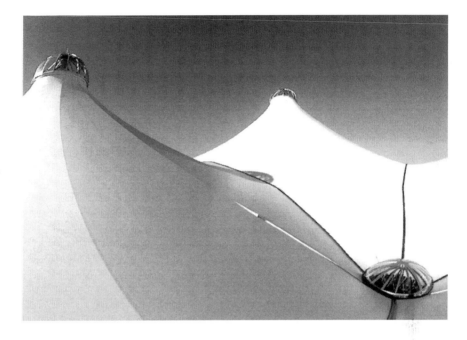

The tie-down point of the inverted tent element is covered with a plexiglass dome. Openings around the ring allow water to drain into a reflection pool.
[6.4.06]

Below the tie-down point a net of steel cables anchors the roof down.
[6.4.07]

to the roof of my rental car and took a photograph. I liked what I saw, even the prefab hemispherical skylights on the tops of the masts whose selection, I felt, had been a weak compromise.

The space inside exceeded my expectations: the floating volumes of the membranes suspended between soft earth berms with no visible sub-structure seemed to have no boundaries. It was more like a sky above a landscape. The abundance of diffuse daylight pierced by rays of sun from the mast-top skylights, strengthened this sense of being outdoors. Trees and plants flourished. People seemed to feel good in this space.

The reversed tent module and its anchorage was a particularly successful feature: terminating in a circular ring, the membrane was held down by a grid of diagonal cables forming a hyperboloid funnel not unlike the shape of concrete cooling towers. The ground anchorage framed a small pool, which would receive the rain water entering through openings in the ring on top.

Watching people in a thunderstorm was fascinating: they would hear the drumming of the rain on the fabric roof; they would see the water gush into the little pool; and they would feel protected, safe and comfortable under the huge, brilliant umbrella overhead.

Later I designed quite a number of inverted roof

structures. The cover for Architect William Morgan's bandstand in Jacksonville, Florida, demonstrates this form eloquently. Two masts extend upward to support the horizontal struts which form the upper supports of the inverted tent units.

My roof design for a tennis condominium for *Innisbrook*, a vacation community near St. Petersburg, Florida, was the most ambitious application. Architect Bob Fessler had designed two interior courtyards surrounded by six-story condominium buildings. The courtyards contained tennis courts, which could be seen from the windows and balconies of the apartments. Inverted tent units were the perfect solution for covering the tennis courtyards. The upper edges of the tent units were attached continuously to the roof slabs of the condominium buildings which would function as the ring beam, restraining the tensile structure forces. Their funnel-shaped downward peaks were anchored between courts on line with the nets.

This arrangement would minimize any external loads the structures would have to carry. There is, of course, no snow in southern Florida. Hurricane level wind loads would cause upward suction, but the particular shape of this structure would shed the wind stream with little impact on the roof surface.

Most significantly, the fabric roof cover over the courtyards would improve the comfort conditions in the building and save energy. The fabric cover would reduce heat gain by reflecting most of the heat and by shading

The bandstand cover in Jacksonville, Florida, has two inverted tent units suspended from an array of masts and struts. [6.4.08]

Model (below) and building plan and section (opposite page) illustrate the courtyard enclosure of the Innisbrook tennis condominiums in St. Petersburg, Florida. The four peaks are above the dining and atrium areas.

The 120 ft. wide inverted tent units provide space for four tennis courts in each courtyard, and create an intermediate environment between the Florida heat and the air-conditioned apartments. [6.4.09; 6.4.10]

Big Tops to Stay

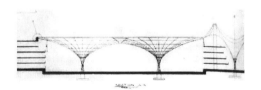

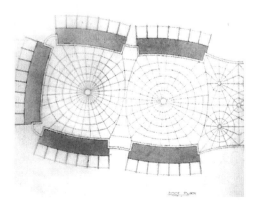

The trussed masts rise among the palm trees inside the African Pavilion at Ashboro Zoo. [6.4.11]

the surfaces of the buildings and the ground inside the courtyards. The temperature in the covered space would end up being half way between the air temperature outside and the cooled air inside the apartments. Night radiation would help to further cool down the space. In combination, this would make it quite comfortable for playing tennis in the courtyards and it would substantially reduce the heat gain of the apartment spaces that faced in this direction. Neither this project nor an even larger version of it planned for a site near Washington, D.C. a few years later, could be financed at the time

Later I proposed this environmental concept for a cover over five city blocks of Saginaw Street (the main street) in Flint, Michigan. The idea was to counter the desertion of downtown commercial activity by turning the main street into a covered mall and thereby help revitalize the city. This, like so many other conceptual proposals, was not carried out because the mayor could not get the financial support from Washington which he had counted on.

The African Pavilion of the Ashboro Zoo in North Carolina, however, was built and has served well for over a decade. Three pole-supported radial tents cover the pavilion which sits on a steep slope. The open-web trussed masts keep the space visually unobstructed. The bottom detail is designed to permit the structure to be stressed by jacking the masts upward, a feature we had first used in Florida Festival. High translucency again supports a rich plant life which gives the space an exotic feeling.

The largest tent modules built to date are the two units which make up the stage cover for the Shoreline Amphitheater in Mountain View, California, not far from San Francisco. It covers an area of 80 000 sq. ft. But this is not a radial tent design. Completed in 1987, this structure has an absolute minimum of cable lines which form the boundaries of four point surfaces. The most prominent feature is the edge catenary in front of the stage, acting like a huge proscenium arch which spans 131m (430ft) from anchor to anchor. Again, the mast bottoms were detailed to allow jacking from the bottom which is so much more accessible than a platform at the mast top.

This structure was initially built with a Silicon coated fiberglass fabric. This had the advantage of having a higher translucency and a lower cost than Teflon coated fiberglass. Since silicon is not a thermo-plastic material, its seams cannot be welded. They have to be glued, and this requires strict environmental control for the gluing process. Because of the extremely tight schedule, the fabricator chose to make a field seam along the ridge line of the of the structure rather than clamping it to the ridge cable. Because of a week portion in this seam, it developed a rip, which then propagating into the body of the fabric membrane.

I interrupted a vacation trip to fly to California and

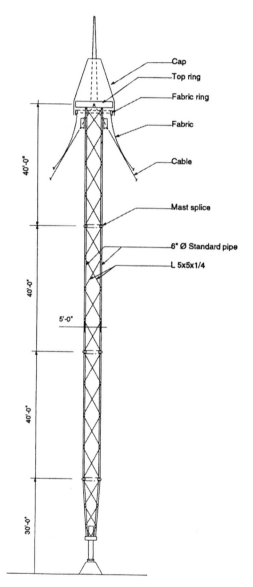

The elevation of the mast shows its dimensions and steel elements. [6.04.14]

The bottom detail (opposite page) indicates the jacking device for the tent and the cylindrical bearing which allows it to rotate. [6.4.15]

Big Tops to Stay

Workers on the Shoreline Amphitheater roof complete the clamping. The distinct fabric panels are clearly visible. [6.4.16]

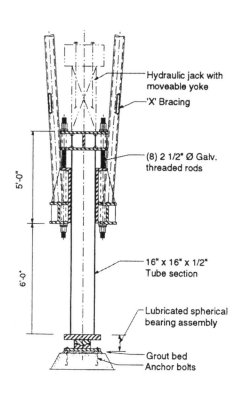

see what could be done. I will never forget this most awful Sunday afternoon, under a gorgeous California sun, watching the afternoon breeze rip the structure to pieces. The good part of the story is that on that Sunday evening I was able to reach by phone all those in the industry that I needed to reach. On Tuesday, all involved parties met in my office in New York where my associate Mark Schlogel had the new patterning coming off the computer. Birdair had located enough Teflon-coated material of sufficient strength and translucency which had been coated for another project. They sprang into action, sending Jim Ford, their most capable and experienced construction supervisor, into the field. He is an engineer with a solid theoretical background and an excellent feel for the behavior of structures.

To watch him in operation is a delight: totally calm, gentle, soft spoken, he is in total control at every step. Nothing is done before all consequences have been considered and everyone knows exactly what he is expected to do. So the project moves forward, quietly and deliberately. And though nothing much seems to happen, everything gets done right and on schedule.

Within 24 days of that awful Sunday the structure was replaced, in time for Madonna to open the season under a new Teflon coated roof.

The largest, most recent, and in many ways, most

6.5 Moving the Masts Outside

The fabric of the Queeny Park structure in St.Louis County is held up by a ridge cable suspended from two A-frames. [6.5.02]

Hanging from its 33.5 m (110 ft) high A-frame, the Queeny Park pavilion roof soars upwards like a large bird. [6.5.01]

The world of structures can be divided into two categories: end-supported and center-supported systems. End-supported structures span clear across a space from one end to the other. Beams, arches, and domes are end-supported; even air-supported structures fall into this category.

Tents, cable-supported bridges and double cantilevers require interior supports, often more than one. A suspension bridge most commonly has two towers over which the main suspension cables are draped. A circus is not only known for its number of rings but also for its number of masts. Tensile structures, with few exceptions, require interior supports.

Architectural spaces often require column-free spans, especially when used for public assembly. One simple way of removing the obstacle of a supporting mast in a tent-shaped roof is by splitting it in two and spreading it apart at the bottom, thereby turning it into an A-frame. At the top the two legs of the A-frame meet just over the peak of the tent. At the bottom they are located at the periphery of the structure. Because of the generic inward curvature of tensile membranes such an A-frame is entirely outside the roof surface.

The roof cover for an outdoor facility at Queeny Park in St. Louis County demonstrates this concept sim-

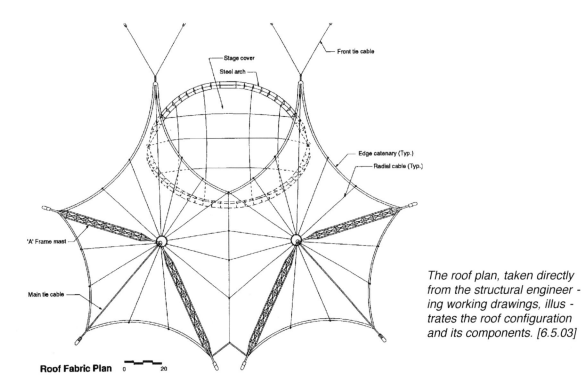

The roof plan, taken directly from the structural engineering working drawings, illustrates the roof configuration and its components. [6.5.03]

ply. The space under the roof is used as a skating rink in the winter and and exhibition tennis court in the summer. Spectator seating is located on two sides. A large and a small A-frame form the roof's main high support points. Tripod frames consisting of pipe columns and cables form the anchors around the periphery of the roof membrane.

The large A-frame is 110ft. high. Suspended from it is the ridge cable which acts as a spine over which the fabric is draped. The opposite end of the ridge cable terminates in a small A-frame, which - leaning outward - "digs its heels in" to resist the powerful pull. Deeply scalloped edge catenaries frame the periphery of the roof. No other interior cables are needed.

The Tennessee Pavilion in Knoxville also uses A-frames as main support elements. The two radial tent roof units covering the auditorium space of the Tennessee Pavilion hang from A-frames which lean inward and are restrained by stay cables. Two cable-braced columns anchor the front edge of this roof, with the edge catenary between these two columns sweeping over the roof of the stage house below it.

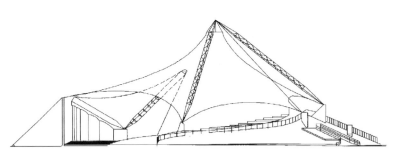

Section through th Tennessee Pavilion showing the stage tucked under the auditorium roof. [6.5.04]

Moving the Masts Outside

The saddle-shaped roof over the stage cover nestles under the auditorium roof of the Tennessee Pavilion. The leaning A-frames and the stay cables which hold them back are clearly visible, along with the radial cables which shape the tent units. The corner tripods, each consisting of a vertical strut and two sloping cables, are connected to concrete anchors rising from the water. [6.5.05]

The saddle-shaped membrane of the stage house roof nestles under the auditorium roof. It is stretched between two inclined trussed arches which rise from common spring points. Vertical stays tie the back arch to the ground. Similar stay cables at the front arch are gathered to an arch-like restraining cable which becomes the proscenium arch between stage house and auditorium. Vertical curtains, attached to the tie cables at the back of the stage, can be raised to open the view to the water behind them for fountain displays and water shows.

The stage house roof, low in the back and high toward the audience, forms an effective soundboard. The tent units over the audience, on the other hand, become "black holes" for sound waves that hit them so that the audience experiences a sense of privacy and intimacy in spite of the unprotected sides which open the pavilion to the noisy world around it.

Built as part of the 1982 World's Fair, the 1500 seat pavilion remained in frequent use for many years with its Teflon-coated fiberglass membrane glowing in the Knoxville night sky. It is in need of repair now and has recently been considered as a historic monument.

The Crown Center Pavilion in Kansas City,

Study model for my Crown Center Pavilion design shows the integrated composition of A-frames and cables which support the roof. [6.5.06]

Missouri, built in 1984, also serves a combination of purposes. In the winter it houses a skating rink; in the summer it is a stage cover for a variety of performances, including symphony concerts and ballet. The design for this structure again uses large and small A-frames as high point supports for the roof, but it takes the integration of the structural system a big step further.

The Crown Center stage had been covered for years with a small fabric roof. Now its owner wanted to replace it with a larger structure which would cover part of the seats in the summer and the entire skating rink in the winter. Selected as principal designer, I was charged with developing a design which would give the Center a new visual focal point but at the same time be functional and economical. One critical restraint was the geometry of the structure below: a multi-level parking garage with columns spaced 30 ft. apart in both directions.

After studying numerous concepts using sketches and study models, I decided to use a combination of four A-frames, two large ones and two small ones. Four piers, positioned 30 ft from two diagonally opposite corners of the structure, form the supports of all four A-frames. They are also anchor points for the edge catenaries and for a set of intersecting valley cables. Two additional piers occupying the corners of the other diagonal were needed to anchor the ridge cables and the other ends of the large edge catenaries. Vertical cables were needed at the peaks of the small A-frames to prevent them from rotating inward.

Ridge cables, suspended from the large A-frames, hold the roof up. Valley cables, extending down to the A-frame support points, hold the roof down. A steel ring

Night performance under the glowing Crown Center roof in Kansas City. [6.5.07]

Moving the Masts Outside

Crown Center Detail: two A-frame legs, two edge catenaries and a valley cable meet on top of this steel pier.
[6.5.08]

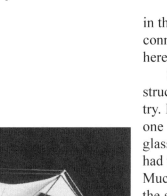

My study model for the integrated fabric shell structure exhibits the components and their relationship: four vertical A-frames, the square formed by the restraining struts, a pair of valley cables and a pair of ridge cables.
[6.5.09]

in the center of the roof is designed to accommodate the connection of all the ridge and valley cables which meet here.

Seemingly complex at first glance, this design is structurally quite simple because of its diagonal symmetry. Built by ODC, a subsidiary of Dow Corning, it was one of the first structures using a silicon-coated fiberglass fabric. Compared with Teflon-coated fiberglass it had the advantage of higher translucency and lower cost. Much of the translucency has remained in the years since the structure was built, but unlike Teflon whose surface retains its silvery shine, the silicon coating turned a streaky grey. This is a problem still to be tackled if Silicon coated fiberglass fabric is to come back into the market.

I had early on pursued the idea of using A-frames more systematically and developed a structure which would integrate an exterior frame and a membrane roof into a system which would balance out all horizontal forces before they reached the ground. It seemed to be a promising system, capable of serving many useful purposes from emergency housing to full size stadium roofs. I even got a U.S. patent (now long expired). It did find a few useful, even exciting applications but never became the marketable product I envisioned, despite sufficient public exposure.

The idea first began with a design which had four A-frames placed in the four faces of a square. Their peaks were connected with a square frame placed on a skew. At this stage - if you build a model - the structure is annoyingly unstable. It will collapse, with two of the A-frames falling outward, two inward. It takes one horizontal strut or two cables forming a cross down the middle to make the structure stable.

The ridge cables of a membrane in tension will serve this stabilizing purpose more gently, letting the A-frames move just a small distance to adjust for each different load case. The ridge cables will also carry the downward loads, such as snow, efficiently. Upward loads, caused by wind suction, can be resisted by the upward curvature of the fabric in the diagonal direction. For larger structures these loads are more effectively handled by valley cables placed diagonally between corner support points of the A-frames. The edges of the membrane can be clamped to the A-frames or to an edge cable located just inside the A-frames. The latter produces the more economical structure; the former may make a more economical building if the A-frames are filled with a solid wall.

The Philadelphia Bicentennial celebration offered the first opportunity of using these modules in quantities. Twelve food pavilions used modules with a 24 ft base, and four mini-stages had a 32ft base. The entrance pavilion to the children's play area demonstrated the versatility of the design by adding side extensions to the base configuration.

For the 1978 Democratic Party Convention in New York City, NBC installed a 32 ft unit with transparent wall panels on a roof opposite Madison Square Garden.

A 32 ft (9.72m) module used by NBC at the 1978 Democratic Convention in New York's Madison Square Garden.
[6.5.10]

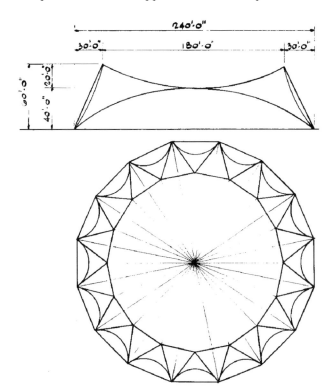

The section and plan drawings of a 240 ft (73.2 m) diameter arena roof illustrates one of many variations of the ten - sioned fabric shell concept.
[6.5.12]

Moving the Masts Outside

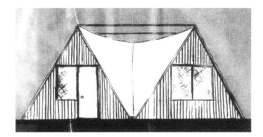

In this sketch I tried to demonstrate how tensioned fabric shell units, erectable in hours after emergencies like an earthquake, could easily be converted into livable housing.
[6.5.11]

The 107 m (350 ft) dia. arena for San Cristobal, Venezuela, with a retractable center, hangs from a ring made of precast concrete elements.
[6.5.13]

What exposure! Here it was on national television, every night. But not even NBC anchorman John Chancellor, who sat under it, seemed to notice. Clearly, this structure did not register as architecture, nor as something useful, exciting or beautiful. Yet at the same time, people kept inquiring about the feasibility of fabric designs for country houses.

I tried to interest international relief organizations in using it for emergency housing. The rolls of fabric and bundles of aluminum pipe would take up little space. Thousands of units could be stored in one warehouse. They could be flown to disaster sites and erected in a day. Supplemented by a small utility box containing the critical elements of kitchen and bath and with the addition of rigid wall units, they could quickly and easily be converted into livable homes. Because of their light weight and flexible construction they would be absolutely the safest structures in earthquake aftershocks. But no takers.

The design for an arena in San Cristobal, Venezuela, was a chance to apply this principle to a large span with a retractable roof spanning 107 m (350 ft.) In this design 24 A-frames of precast concrete connect to a top support ring. The peripheral ring of tents suspended from this

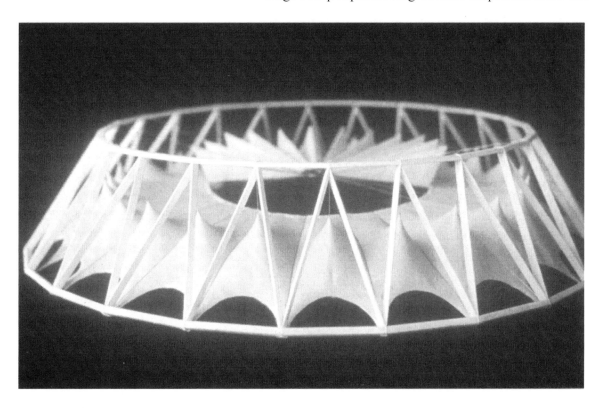

The picnic pavilion at Sea World in San Diego extends this 10-sided tensioned fabric shell by adding cantilevered sections. [6.5.14]

structure would shield the spectators against the fierce sun. The fan-shaped elements of the center portion, hung independently, were removable, permitting adjustment for various events and weather conditions.

On this project communication with the architect in Venezuela was particularly strange: A certain Mrs. Varga, whose husband was in a New York hospital, would act as messenger, carrying drawings and models, which we handed her in exchange for certified bank checks. In the end, we never heard what happened. Even the telephone number in Venezuela went dead. It was one of the many designs which were not built.

Of the many proposed designs using this principle, two large structures were actually built. The first was a picnic pavilion for Sea World in San Diego, built in 1980. It further extended the concept by adding cantilever frames to the ten-sided structure. A few years later it was enclosed in glass, which destroyed much of its original charm.

The most significant application was the design for a Bullock's department store in San Mateo, California. We had worked with Bullocks of Northern California and its enthusiastic chairman, Paul Heidrich, for years and had put an arch-supported fabric roof on top of his store in San Jose in 1978. He now wanted to use tensile architecture for this new facility, for the excitement that

Moving the Masts Outside

daylight would bring to the space, and for the substantial savings in energy.

Initially, both the architect and the store designers were opposed to the octagon design I proposed. Being the structural consultant I was the tail wagging the dog. But Paul liked it. And once the store designers fell in love with the rotational layout - which would keep customers circling and, they hoped, buying - the battle was won.

An eight-sided ring, housing offices and storage spaces, surrounds the 296 ft diameter store space which the fabric roof covers. Eight concrete columns inside this space support A-frames from which the folded membrane is suspended. Stay cables tie the A-frames back to the flat concrete roof which forms a solid ring around the tensile structure, eliminating the need for a compression ring at the top of the A-frames, making the structure simpler and more elegant.

The light inside the store was glorious. Part of its charm came from the partial inner liner with its butterfly wing shapes. Strangely enough, the design of liner configuration was the consequence of the fire marshall's decision to require an inner liner only in areas where the

The former Bullocks department store in San Mateo, California, is the largest application of a tensioned fabric shell. 8 A-frames, rising from interior concrete piers, form high and low points. The rigid peripheral roof acts as the restraining ring. [6.5.15]

main fabric surface had less than 20% slope. He derived that logic from the results of the burning brand tests. This fire resistance test requires that a 9" x 9" x 2.5" block of wood is placed burning on the roof surface, and is kept alight by a heavy blast of air until totally consumed. Not a spark is allowed to fall through the roof in this process.

A single skin of Teflon-coated fiberglass satisfied this requirement in half the tests. A double-skin would always pass the test. Assuming that the burning brand - a branch from a forest fire, for example - would slide off the sloped portion of the building, a double skin was only needed in the flat areas.

Satisfying this requirement with the curvilinear vocabulary of tensile architecture enriched the opulent space of the store. From the air its peaks can be seen just before landing towards the north at San Francisco airport.

Time Magazine selected the building as one of the ten best designs of 1981.

A-frame supported tensile structures invite an almost unlimited variety of design configurations. A few are shown in this chapter and on the next page page.

The interior of the Bullocks department store. The partial inner liner - basically complying with the requirements of the fire marshall - adds to the charm of the light-filled space. [6.5.16]

Moving the Masts Outside

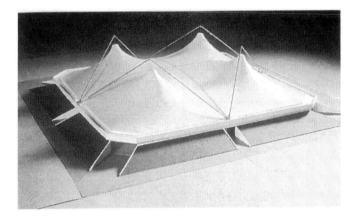

Four A-frames support the roof of an open-plan school project in Jacksonville, Florida, by architect William Morgan. [6.5.17]

The simplest application is for a music pavilion in Eilat, Israel, where the roof hangs from just one simple A-frame.

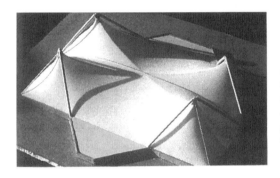

In this design for a sports arena in St. Louis County, Mo., the A-frames are stretched apart to form support points for diagonal ridge cables. [6.5.18]

A single A-frame holds up the roof of this 1500 seat auditorium in Eilat, Israel, completed in 1994. [6.5.19]

The huge fabric umbrella protects and shapes the spectator space of the King Fadh Stadium in Riyadh, Saudi Arabia.. It reflects the heat of the sun, filters its light, dissipates sound and exudes a sense of intimacy that defies the huge scale of the building. [6.6.01]

6.6
A Giant Flower Shades the Stadium

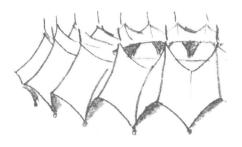

Variations of distance and position make each of the 24 units look different in this giant circle of tents.
[6.6.02]

The most welcome response to my work came from a young architect when she said: "I like your structures because they are so human." Although this was the response I had hoped for, I had never really been aware that I was shaping my structures with any particular effect in mind. I usually had a clear image of the space which, I thought, would best serve the people who would use it. The form of the building then grew out of the process of developing the structure which would most naturally create this space. In this process I was guided by the discipline of the structural order as I understood it. I looked for clarity and simplicity, eliminating what was not needed, emphasizing what was there. And I always paid attention to scale and proportion - that part came natural.

Looking back, I understand that this process worked well for my designs, and rather than being coldly objective, it expressed my personal outlook It worked particularly well for the largest span I designed: the roof of the King Fahd Stadium in Riyadh, Saudi Arabia. In spite of its huge scale, the space under this roof has a sense of intimacy not commonly found in large spectator facilities.

The initial design was a large-scale application of a tensioned fabric shell, suspended from a braced A-frame ring. An extensive shade area was desirable to reduce the heat of the desert climate. But the playing field was to be open to the sky to meet the requirements for World Cup football (soccer) and for an Olympic stadium. The circular roof over part of the oval stadium, the 65,000 spectator seats, and the ample concourse areas covered an area of 560,000 sq.ft. (Not counting the 134 m (440 ft.) diameter center opening - which could have easily been closed with little change to the design of the structure - this was 1.4 times the area of the Astrodome.

I produced the design in a Kensington hotel room. We had arrived in London on a Tuesday morning in June of 1980, and had gone straight to the office of Fraser Roberts, the architects who had pursued this project for years. A design put out to bid a few years earlier had failed to meet the budget by a large margin, at least in part because of the cost of the arch-supported cable net roof which, like the Munich Stadium roof, was to be covered with clear acrylic panels. The success of the Jeddah airport roof had helped persuade the Saudi authorities to try again, using fabric tensile technology. When Ron Mc Clochlan of Owens Corning Fiberglass, Dr. Suko of NKK, the prominent Japanese steel contractor, and I all met with Ian Fraser that morning he told us that he needed a design by the next day: he was scheduled to leave for Riyadh on Friday and had an artist waiting to do a rendering.

Because they were still in the process of moving into this office there was no desk for me to work on, Ian Fraser told me. But he could provide me with some drafting tools and paper. Yes, he would see to it that there was coffee the next morning when we were to return with our design.

From twelve noon to twelve midnight I made three drawings, while Ron Mc Clochlan and Dr. Suko, with long distance help from their offices, worked on cost estimates and construction schedules. At eight o'clock the next morning we delivered the drawings together with budget and fee proposals. Ian Fraser was delighted. We sipped our coffee and booked our flights back to New York, Toledo, Ohio, and Osaka.

Exterior view (above), roof and seating plan (below)' and typical section (bottom) from the initial design drawings which I made in a London hotel. The initial concept was a ring with 24 tent units supported by a braced A-frame ring structure. [6.6.03;6.6.04;6.6.05]

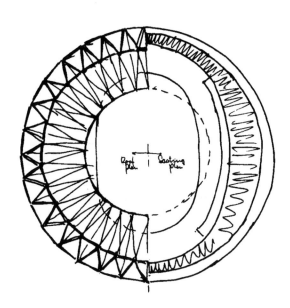

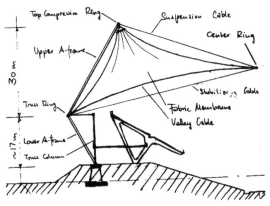

A Giant Flower for the Stadium

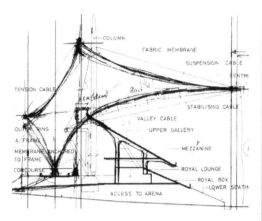

Overlay sketch from the design process: the final shape of the roof units begins to evolve. [6.6.06]

A partial stretch fabric model served to find the optimal lines for patterning. The lines were drawn on masking tape. [6.6.08]

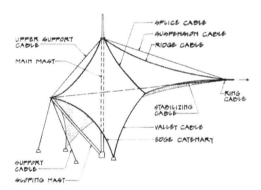

Fabric membrane is stretched between ridge cables, valley cables and edge catenaries. Ridge cables and masts hold the system up. Valley cables and bottom anchors hold the system down. The sloping mast and edge tie-downs hold it outwards, and the ring cable holds all horizontal forces in balance inward. Cable network and fabric membrane duplicate force nets for redundancy. [6.6.07]

There was only one primary change to this design in the revised version we delivered a month later in response to the comments by the Saudi Ministry of Health and Youth Welfare. They wanted the roof more tent-like, Ian Fraser had reported. This signified a sea change in attitude since the days of the Jeddah design when the Saudi authorities would have preferred "Western architecture"- meaning conventional steel or concrete construction. They had now come to appreciate the symbolic link of these new tent forms to the tents with their own ancient nomadic past.

It was easy to comply with this request. We replaced the A-frame system with straight, vertical pylons under each of the tent peaks. We retained the shape of the fabric membrane, except that in the back we brought it down to anchor points on the concourse floor. A sloping mast and tie-down cables added the vital peripheral anchor. The structure became simpler, cheaper, and much clearer in its form. The interior masts, we found, encroached into the seating area only in four places at each of the narrow ends of the oval. Except for some fine-tuning, the concept of the roof and its form were set.

The roof is a circular ring consisting of 24 identical units. Each unit is a tent form with a single peak supported by an 58 m (190 ft) high tubular mast. The masts sit on a circle of 246 m (807 ft) diameter, large enough

The finished structure looks more precise and beautiful than computer drawings and study models. [6.6.09]

to fit any existing stadium, - the Houston Astrodome or the Pontiac Silverdome - inside with room to spare.

Each unit has four fabric elements: a front and back panel and their mirror images. They span between ridge and valley cables in one direction, an edge catenary and a radial splice cable in the other. The valley cables and the outer edge catenaries meet at an anchor point halfway between mast footings. Small reflecting pools surround these anchors and act as overflow basins for the drains, need to carry the rainwater from the roof in the rare, but violent, downpours which are characteristic of this desert climate.

The structural function of the ridge and valley cables is duplicated by suspension and stabilizing cables. This means that the system is stable without the fabric panels which made it possible to erect the mast and cable system first and subsequently install the fabric panels, one at a time. It will also allow the later replacement of fabric panels and gives the structure redundancy and greater

The wind tunnel model shows the whole configu - ration of the entire roof. [6.6.11]

Tensile structures are unifiers. The outside and the inside are but dif - ferent views of the same structural forms. [6.6.10]

A Giant Flower for the Stadium

The 134 m (440 ft) diameter ring cable is the stabilizer of the system. The inner ends of all cables are connected to it, and a catwalk with light fixtures hangs from it. [6.6.12]

safety.

The central ring cable acts as the equalizer of horizontal forces and as moderating unifier of the structural system. The forces in this cable never change much and its largest movements are projected to be less than a meter up and down even under the most violent storm conditions.

The anticipated loads from various wind conditions were determined in wind tunnel tests conducted by Dr. Alan Davenport and his team in the *boundary layer* wind tunnel of the University of Western Ontario, with whom we had worked earlier on the Jeddah Haj terminal project. Critical load distributions from the wind tunnel testing were then used in the non-linear computer analysis to simulate the behavior of the structure, establish the variation of stresses in all its components and predict its deformations. The full scale stress measurements on the Jeddah test modules had established the validity and reliability of our computer simulation, and by this time the process had matured to become fast and efficient. The analysis demonstrated that this giant structure, including its most daring feature, the interior ring cable, would behave excellently under the predicted loads.

This was the largest structural system I had designed and - in my mind - one of the simplest. It had been put on paper faster than any of the other designs, but it also had the longest gestation time, since the ideas which led

Two interior fabric panels are ready for sliding into position along an inclined erection cable. (above) [6.6.13]

Construction workers complete the clamping of adjacent fabric panels along the valley cable of one tent unit. [6.6.14]

Erection workers complete the fabric splice along a ridge cable [6.6.15]

A Giant Flower for the Stadium

The stadium roof seen through the entrance canopy. The workers in the foreground provide a sense of scale. [6.6.16]

to it had been evolving for many years on numerous designs. The design was completed with a trip to Riyadh to present the project and assure government officials of its reliability.

When the green light came for construction, early in 1983, the design/construct process which had been the basis for the design phase of the building was replaced by a bid process. Philip Holtzmann of Germany became the general contractor, Birdair Structures replaced Owens/Corning Fiberglass as the roof contractor. As part of the design/construct team we lost our position in the engineering supervision of the project which was shifted to the contractor. For the first time in my experience, we found ourselves on the sidelines for most of the construction process. It was fortunate that the contractor hired Schlaich and Partner (now Schlaich, Bergermann and Partner) as consultants for the construction process. Their contribution resulted in an elegant redesign of several structural details, replacing weldments with state-of-the-art steel castings and applying their special expertise in cable technology to assure corrosion protection for the cable system.

Birdair, who had meanwhile installed its own computer program for shaping and analyzing fabric struc-

140   **Light Structures
Structures of Light**

View of the completed Stadium. [6.6.17]

tures, did the patterning and detailing of the fabric membrane components. My former associate, Steve Goodson, who had joined Birdair was the chief project engineer in their office. Jim Ford was in charge of construction in the field.

Birdair developed an ingenious erection process. As intended in the design, the masts were erected first and the exterior tie-backs were installed. The interior cable ring, consisting of six 58 mm ($2^3/_8$ in) locked coil steel cables, was assembled on the ground. Ropes running over sheaves at the mast tops were used to raise it in

A Giant Flower for the Stadium

position. The cable net was then completed and tensioned.

With the supporting strut and cable system in place, the fabric elements were laid out on the ground. They were then attached to two erection cables which ran up to the top of the masts. Not unlike a cable car, each fabric element was then pulled up along these cables into its proper position in the structure and was clamped to the structure's cable system.

It was only at the very end of the construction process that I had an opportunity to visit the site and review the structure. Today, after almost twenty years, it is still the largest span roof and covers more area than any other stadium roof.

Because of the simplicity of its form and the direct flow of forces from the location of the load to anchors and foundations, it is a highly economical way of covering a very large space. It makes a wonderful space without clutter of structural elements. The play of sunlight and shadows on the deep fabric folds keeps the space alive and vibrant. Sound is dissipated, not focused as in most dome-like structures. Water from snow and rain has a clear path down the valleys, avoiding the danger of ponding.

The huge assembly of simple white tent shapes rises from the limitless desert into the dark blue sky, a joyful affirmation of the power and beauty of the physical order of this world we are permitted to live in for some while.

Intersecting trussed arches span tennis courts in a suburb of Atlanta, Ga. The fabrics anti-clastic curvature is clearly visible. The 30% translucency helps to "see the ball". This early version of silicone-coated fiberglass has turned grey on the outside, but retained the translucency [6.7.01]

6.7
Stretched Over Slender Arches

The mechanism which enables a structure to span space requires the paired combination of tension and compression elements. There is no exception to this rule. But it is not necessary for both types of elements to cross the span. Parts of the structural system may not even be apparent. An arch, rising from the ground, is a classic case in: the arch transmits load directly to the support points. The tension element of the total structural system may be a tie beam under ground, or the foundation elements may transfer the thrust from the arch into the rock

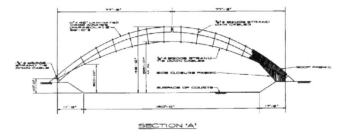

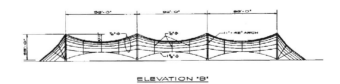

This 1969 design for a 6-court tennis structure uses a cable reinforced membrane supported by rigid arches spaced 96 ft.(29.28m) apart. The arches rest on an earth berm to eliminate the need for exterior walls, give the space extra height, and help the foundations to resist the thrust of the arches. [6.7.02]

or soil it sits on, replacing the need for a tension element.

When arches are part of a dome, the tension element is a peripheral ring at the base of those arches. Because it appears simply as the top edge member of the enclosing wall, a casual observer may not associate such a ring with the structural function of the dome.

Arches and domes by their nature are end-supported. They are high in the middle and low at the edges, which makes them well suited for certain architectural spaces such as spectator-sport facilities. The arch, therefore, has its place as a support element in fabric tensile structures, especially when the arch elements can be shop fabricated, transported on conventional vehicles and erected without exceptional shoring and lifting equipment.

Because of my past work in concrete structures, thinking of arch and dome forms came naturally to me. Indeed, one of my first fabric structure concepts was a dome-like structure. Its viability was first tested by a 30 ft. diameter octagon design fabricated and erected by architecture students at Columbia University for Commencement Day of 1972. To avoid the cost of fabricating arches to the proper curvature, the arches were made of flexible aluminum bands, 75 mm wide and only 19 mm thick (3" x 0.75"). They were supported by eight A-frames arranged around the periphery. A ring cable balanced the horizontal thrust from the arches at their spring points. The membrane contained valley cables in its deep folds which helped stiffen the structure and balance its internal forces.

It took less than a day to erect, and it demonstrated the simplicity of this design. But when I came with my VW bus, a few days later, to remove the structure, I found it on the ground in a tangled heap of fabric and bent aluminum. Boys from the neighborhood had used it as a large trampoline, the guard at the library explained. On the third day, they had walked along the top of one arch, three in a row. It flipped through, causing the dome to collapse, much like an umbrella in a gusty wind. Fortunately, the boys were not injured, but the in event demonstrated the fallacy of flexible arches.

The reason for the failure is simple: while cables under load take on a shape which can carry greater loads, arches flatten. This causes the force to increase and, being compression members, they buckle.

For the next arch-supported fabric roof design I

In this flexible-arch fabric dome, the structure's main tension element is the barely visible ring cable. This experimental structure was erected by architectural students at Columbia University under my guidance, in 1972. [6.7.03]

Stretched over Slender Arches

Laminated wood arches support the Teflon-coated fiberglass roof of the Bullocks department store in San Jose, CA. [6.7.04]

Hydraulic jacks are used to stress the fabric, a method I adopted from post-tensioned concrete construction. [6.7.05] (above)

Triangular trussed arches, shop fabricated of stgeel pipe, are economical components for fabric domes and vaults. [6.7.06]

The 138 ft span arches of St.Mary's field house uses steel angles for the web members. [6.7.07]

chose rigid arches. Spanning 33.6 m (96 ft.), the roof for the Bullock's Department Store in San Jose, California had arches of laminated wood construction which intersected in the center. Fabricated and shipped in halves, they formed a stable frame once three sections were connected in the center. This cross-arch design became the basis for numerous roof designs using steel and concrete arch members. A covered tennis court in an Atlanta suburb was the next application. Its truss-like arches had a triangular cross section with 5" (125 mm) diameter steel pipes forming the three-chord members. The circular curvature of the arches was achieved by cold-deforming the pipes to a constant radius using the same equipment for all of them. Pipes of 1" (25 mm) diameter were then cut and welded into position to form the web members. Because of the much larger diameter of the chord pipes, straight saw cuts were sufficient to achieve a good

weld.

In other designs small gusset plates were welded to the chords. The web members were then cut square and provided with a slot. They were slid over the gusset plate and connected with fillet welds. In another variation the web members were made of steel angles, which cost less than pipes. These methods require less accuracy of fabrication and simplify the fabrication.

Pipes and tubes up to 180 mm (7") are easy to curve without heating. For U.S. applications, arch sections just under 40 ft. length are particularly economical, since this is a common mill length for pipe material. They are easy to ship, handle and assemble. Several of these sections can be bolted together on site to form larger arch elements, usually making up half the total arch. A special connector in mid-span, either placed on an erection tower or connected to one of the half arches, provides the means by which the arch system is installed. Cross arch systems of this type were used in a large number of our designs, serving many different purposes from gymnasiums to warehouse and bandstand covers.

The largest arch-supported roof using this steel arch system is the University of Wisconsin's McClain practice facility in Madison, which mainly serves the

The 67 m (220 ft) long steel arches of McClain during erection. A py-wood decking makes them accessable to the workers. [6.7.08]

Interior of the McClain football practice facility of the University of Wisconsin. The center protion of the roof is transluceent. [6.7.09]

Stretched over Slender Arches

Computer perspective of the Wimbledon practice facility. This 1986 image was produced with our own software. The net shown was used for the design and analysis of the structure. [6.7.10]

University's football team. The upper level of the building contains a football practice field covered by a clear span roof of 67 x 101 m (220 ft by 330 ft.). This structure has a number of special features: the arches are parallel; only the center half of the roof is covered with fabric; the two outer quarters are opaque. The roof skin of these sections is a stainless steel roofing system, supported by metal deck on steel truss framing.

This arrangement had many benefits. Restricting the translucent roof to the center half produced the best possible light conditions by maximizing overhead illumination. Lamps hung from the solid roof sections light up the underside of the fabric roof, simulating the daylight condition at night by capitalizing on the high reflectivity of the Teflon surfaces. Heavy insulation of the opaque roof sections reduces the heat losses in winter nights while the low heat absorption and high reflectivity of the Teflon skin minimizes heat gain from the hot summer sun. Mike Dittka, the well-known manager of the Chicago Bears football team, called this the best indoor space his team has ever practiced in.

The enclosure design of the indoor practice facility for the AELTC (All England Lawn Tennis and Croquet Club) at Wimbledon is based on three sets of cross arches. To keep the space free of the "visual noise" often generated by structural components and their dominating geometries, we placed the arches on the outside and the roof hung from them. Earlier experience suggested concrete to be the economical choice, requiring less maintenance in an exposed structure.

With the supporting structure on the outside, the fabric could be assembled on the ground and raised, thus avoiding one of the shortcomings of arch-supported systems with internal arches. They require that the fabric membrane to be lifted over the arches and unfurled from that awkward position.

My guide in this design was the successful site-precasting of arch and shell elements of the concrete dome which I had designed for the University of Virginia field house (1965). But every project is different. Instead of making forms at the site (in Virginia earth molds were used) the contractor chose to use one of his precast plants for fabrication of the arches. The plant was located at the other end of London. So the transportation of the 24.4 m (80 ft) long concrete arches - components of a "light-weight" roof structure - were reportedly the

148 Light Structures
Structures of Light

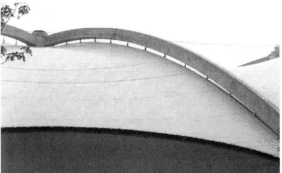

heaviest loads ever to be driven through the streets of London!

After placing the half-arches on their pier supports at the outer ends and a temporary erection tower in the center, the steel extensions were connected with bolts. Cast-in-place keystone sections in the center and "knuckles" at the piers merged the various elements into a continuous concrete frame. The two vinyl coated polyester fabric segments were laid out on the ground, spliced along an arch line, and pulled up by winches located on the arches. After attaching the entire edges - to catenaries on three sides and a continuous curb on the fourth - the membrane was stressed in stages by raising the suspension cables, which ran through the arches in radially oriented sleeves. Closure panels filled the space between edge catenaries and the rigid masonry wall as integral parts of the tensile structure system.

Inside, my objective - shared with Ian King, the project's architect - had been achieved: the roof has become a sky. It transmits diffuse natural light in abundance yet unobtrusively, and exterior is elegant but unassuming. Standing across the street from the main compound of the famous tennis club, you must know what you are looking for to find it during the championship.

Left: The precast concrete arches are in place, ready to receive the fabric which will hang below. [6.7.11]
Above: Steel cable hangers connect the membrane to the arches. Catenaries form the edges of the light main roof and the dark closure panels. [6.7.12]

The completed structure. During trhe Championships, it is usually visible behind courts # 3 and #13. [6.7.13]

Stretched over Slender Arches

On the inside the fabric roof is a "sky" showing no structural members. [6.7.14]

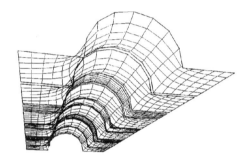

Computer drawn perspective. [6.7.15]

This Wimbledon project began years earlier with proposals for a retractable roof over Center Court. The work for the AELTC was part of a consulting activity which brought me to London several times a year for more than a decade. Another result of this activity was the ceiling structure of the Great Hall of Alexandra Palace in the north of London.

It had started as part of a competition of roof designs to replace the structure which had burned down a few years earlier. We were competing against a space truss design. Though I thought we had won that competition, neither scheme was adopted. The final design solution was a combination of a truss structure with a glass roof cover and a fabric inner liner. The interior columns of the 1865 roof structure were removed, but the interior profile of the old ceiling was kept. A special Silicone-coated lightweight fabric with a 90% translucency was fab-

The ceiling of the Great Hall of Alexandra Palace in London has a translucency of almost 90%, revealing the glass covered trusses above. [6.7.16]

Arch supported silicone-coated roof units at Callaway Garden's Sibley Horticulture Center transmit 50% daylight. [6.7.17]

ricated and installed by Koit of Germany (which also installed the Wimbledon roof and several of our designs later on).

Although arches, especially in combination with cables, still prove to remain the most powerful means of covering super-large spans, they also have a place in short-span structures. The Sibley Horticultural Center at Callaway Gardens in Georgia, US, designed by the architect Kirk Craig and the landscape architect Robert Marvin, is one of the most beautiful projects I have been involved in. The fabric elements supported by cross-arches, merge with the lacy, rust colored steel frame into an intriguing interplay of light, nature and structure. Square modules of 60 ft. (18.3 m) and 30 ft. (9.15 m) side length carry highly translucent fabric panels. Diffuse light from these roof panels mixing with light from clere-story windows, glass doors and glass block walls allow the growth of even delicate plants and give the building its atmosphere of sparkling delight.

*Facing page:
The fabric units shine bright in the sparkling scene of water, plants, and lacy steel supports.
[6.7.18]*

*In the proposed roof cover for the Bayamon baseball stadium in Puerto Rico, the trussed steel arch concept is extended to a span of 168 m (550 ft). The cable reinforced fabric hangs below the steel arch. The arch serves for lighting and ventilation.
[6.7.19]*

Stretched over Slender Arches

The Cynthia Woods Mitchell Center of the Performing Arts, located at The Woodlands, a new town 45 miles from downtown Houston, Texas. It was built for 3000 fixed seats under the roof and 7000 seats on the lawn. Light truss members form A-frames, columns, and lateral struts. Follow spot platforms project from columns, and loud speakers piggy-back on tie-downs. [6.8.01]

6.8
Music under the Tent

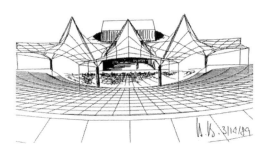

This 1989 perspective, incorporating computer graphics from the design process for the Mitchell Performing Arts Center, illustrates the relationship of stage, auditorium space, lawn area, and roof structure which brings them all together. [6.8.02]

Not unlike Santa Claus, I entered the design process of the open air performing arts center at "The Woodlands" via the roof. The Woodlands is a planned New Town forty-five miles from the center of Houston, Texas. Now (2005) in its fourth decade, it has a population of over 180,000 and is steadily growing. It is the dream child of George Mitchell, Texas energy entrepreneur, philanthropist, and man of inexhaustible imagination. He has revitalized his home town, Galveston, Texas; and The Woodlands is easily the most successful planned community in the USA.

In 1984, Mitchell was looking for an exciting roof structure which would give the schematic design for his performing arts center the feature which would bring it together and give it the signature it lacked. The center was to become the summer home of the Houston Symphony. But it also was to support a wide range of other performance functions, including musicals, ballet, rock concerts, popular singers, and even plays and opera. The Woodland's senior vice president for planning, Richard Browne, remembered the roof design for the Interama project where he had been involved as a civil engineer and planner. He asked me to come to The Woodlands to see what could be done. It was the first step on a path with many ups and downs.

This path began with the exploration of tensile roof forms for the existing design scheme, a process that led

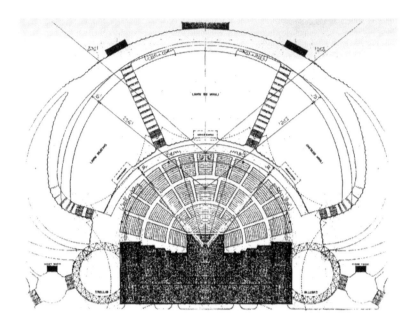

The large 1984 design had 5000 fixed and 7000 lawn seats, a 90 ft stage loft, and a 90,000 sq.ft. roof. The plan drawing shows the main elements. [6.8.03]

nowhere; my designs would not fit into their scheme and I did not fit into their team. I resigned.

George Mitchell's response was to ask me to go beyond the design of the roof and to critically study the various components and look at the design as a whole. Some improvements were obvious: large retaining walls and huge flights of stairs could be replaced by simply shaping earth; other elements could be made simpler, clearer, more efficient. Step by step a new design began to take shape.

Chris Jaffe, the well-known acoustics consultant and John Toffoli, the theater consultant and manager of the Concord Pavilion in Oakland, already working for The Woodlands, became key members of my design team. They and others helped me appreciate what makes this kind of facility work: sight, sound, access, atmosphere, and a clear relationship of all these critical components to each other. I started the design by gathering the seating closely around the stage in a wide arc and setting it into a bowl for best sight lines was. I expanded the bowl into the grassy mound to be built behind the fixed seats, dishing it gently, shaping the earth like a potter, and finished it by laying the bowed ribbons of the paths into its slopes, and adding the lobby circles like curls in a brook. Structure was minimized because the sculpted earth served the necessary functions naturally and obviously. Stairs and slabs for the fixed seats were embedded. Only the stage house and the ancillary facilities, concession stands and toilets, rose up straight and rigid.

Music Under the Tent

The sharp peaks and gentle folds float over the softly molded earth. The study model shows the appealing features of this design. [6.8.04]

Above, stretched between filigree frames like a giant spider web, the white peaks of the roof, edged by scalloped cables, embraced it all. The roof's light fabric swoop complemented the solid mold of the earth, while slender columns, placed to organize and orient the auditorium space, allowed the sides to stay wide open.

The idea was that daylight entering through open sides and filtering through the translucent roof of the completed building would make the inside bright. Nature, earth, and structure would come together productively and joyously. This process of integrating, simplifying, and minimizing hard construction had the additional benefit of reducing the projected construction costs and making the project economically feasible. Our design team was selected for final the design.

The 1984 design was generous: the 90,000 sq. ft. roof floated over the 5,000 seats, the stage house, and part of the adjacent "lobby circles". Space for a further 7000 on the lawn swelled the audience capacity to 12,000. The two large lobby circles, with circular fabric arcades and access to concession stands and toilets, were able to serve the entire audience before the show and during intermissions. The five single-peak segments of the roof were hung from trussed steel A-frames, supported by concrete piers strategically located in line with the radial walkways. As the design model demonstrated, it all came together beautifully.

The budget estimate was only $13.5 million, substantially less than what comparable facilities had cost to

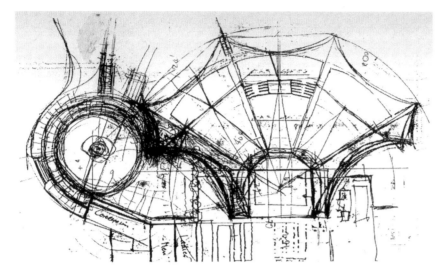

The 1989 design retained most of the features from the 1984 scheme, although everything was smaller and more efficient to keep within a much tighter budget. My overlay sketch shows the outline of the building.
[6.8.05]

build. (Wolf Trap, for instance, had cost twice as much). But it was not to be; at least not at that time. Oil prices plummeted and a recession set in, with the result that the project was shelved.

It was in December of 1988 - almost five years later - that Richard Browne called and asked for ideas for a pilot project, with a temporary stage and rented chairs, (1500 of them). "Mr. Mitchell is willing to invest up to $1 million", he said. "Just to get started, to test the idea, to keep it alive."

Every time we met to present the progress of the design, it grew. By April of 1989 the project had expanded to 3000 fixed seats and 7000 places on the lawn. A fabric roof of 28,000 sq. ft., covered the fixed seating. My proposals to make it larger for future expansion were shot down by Woodlands officials who kept a keen eye on the budget.

From the earlier design the project retained the composition of the major elements: the dish-shaped curvature of the seating and lawn area; the lobby circles; and the A-frame supported fabric roof construction. The roof was reduced to three peaks, however, and integrated into the steel framed stage house. The stage loft was reduced from 90 ft. to a 56 ft. height (which would permit the raising of 24 ft. high scenery). The lobby circles were carried out as a simple wood structures.

Almost everything was built as designed - sometimes not without lengthy battles. The most effort was needed to keep the stage height at a compromise level of 48". (The "rock" people wanted 60" - a dimension which would have made ballet and theatre ridiculous.)

By September, the project was under construction and on April 4, 1990 it opened with a performance by the Houston Symphony, followed the next day by Frank Sinatra. The

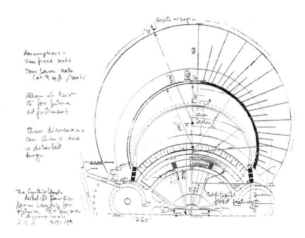

Plan layout, exploring future expansion of the earth berm.
[6.8.06]

Music Under the Tent

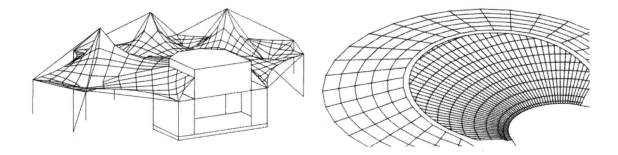

Computer perspectives show the integration of stage house and roof above, [6.8.07] and the shaping of the lawn into a gentle bowl. [6.8.08]

Cynthia Wood Mitchell Center of the Performing Arts has since had over 40 successful performances every summer. Indeed, a few years ago it has been expanded to 5000 fixed seats and 10000 lawn seats.

The smaller roof covers only the fixed seating area. (And it has not been extended over the added fixed seating). It is anchored to the front of the stage house, which has a flat roof of its own. Four trussed steel columns hold up the three A-frames. Horizontal struts link the tops of the columns to the stage house. Other struts reach out from the tops of the columns to the periphery. Together, they neutralize the horizontal forces caused by the tensile forces of the fabric and cables, so that only vertical loads are transmitted through the columns to the foundations. The fabric membrane is held up by its high points under the A-frames, and held down by the low points on top of the main masts. Ridge cables lead from the periphery to the A-frame peaks and back to the stage house. Valley cables scallop from the stage house down to the mast points and back out to the periphery.

Radial isles and walkways are placed in line with the columns to avoid sight line restrictions. Follow spot platforms are attached to the columns; loud speakers for the lawn area are hung from the tie-down members immediately to the rear.

Except for small areas near the tie-downs, all the water follows the valley cables to drains in the main columns. To avoid the danger of the membrane roof filling up with water if a drain is clogged, the intakes are designed as a two-layer funnel with a thin gap between the two layers. In case of a water built-up it can escape inside the space, running down the column to a floor drain.

Sound was, of coarse, a primary concern, especially because of the large volume under the roof. To avoid delayed echoes from the smooth, tensioned ceiling surface which would act like a drum, conventional experience with high ceilings called for sound absorbing soft elements below the ceiling. But the curved and folded surface geometry defeats all conventional experience by dispersing the sound which hits it, letting much of it through, and offering no focus. A remedy in form of sound absorbing inner liners - which I called the "diapers" because of the peculiar shapes we came up with - was never needed. Indeed, the space has good acoustics for rock and classical music alike, in spite of their contrasting characteristics.

My greatest concern was the space itself: whether its proportions would be right; how it would feel to be inside; and how it would appear from the lawn. Nothing could help to accurately predict these aspects, neither drawings nor study models. And in the rush of this final design process, there was never even time to build a good scale model. It remained a question of judgement.

In fact, at the last possible moment in the design process, I decided to fine tune the proportions of the

Lobby circle sketches. I wanted it to be a large sun dial with a delicately detailed wood trellis. [6.8.09]

My trellis design was never built in this form. The much cruder version, which was carried out, is long gone. [6.8.10]

Music Under the Tent

Concert Night. Under the roof the space is intimate as any enclosed concert hall. The people on the lawn are drawn in, with the roof acting as a large, inviting proscenium arch. [6.8.11]

space by lowering the periphery of the entire roof by five feet. As we worked around the clock under heavy time pressure, no one in the office liked this change; but we made it, and it proved to be right.

Sitting under the roof on opening day listening to Rachmaninov's 3rd Piano Concerto, the auditorium space felt as intimate as Carnegie Hall, in spite of a raging thunderstorm. Walking up the lawn area was even more satisfying: the roof with its arching edges framed a festive spectacle of light focussed on the the stage and the performance. My delight was confirmed by a Woodlands dweller saying to his neighbor: "This is great. If I knew it was going to be like this I wouldn't have opposed it."

That was 15 years ago. For years the main complaint had been that too many events are sold out. That's why the facility has been enlarged, adding 2000 fixed seats and 3000 lawn spaces. Unfortunately the roof was not extended to cover the new fixed seats. Yet, the Performing Arts Center at the Woodlands is a continuing success. And the fabric roof looks like new on a bright Texas summer day.

Truss members are being assembled and connected to the stage house to form the roof support system of columns, A-frames, lateral struts and tie-downs.
[6.8.12]

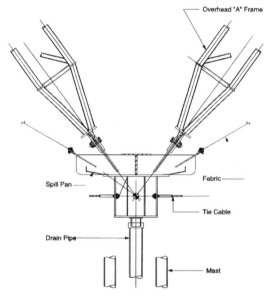

The detail at the top of the columns is shown in the drawing above [6.8.13], and in the photograph on the left. Note he "pan" with its side openings to let water overflow in case the drain gets clogged.
[6.8.14]

Music Under the Tent

161

The finished roof on opening day. [6.8.15]

Ready for the performance. [6.8.16]

The fabric roof of the convention center - like the sails of a giant ship- rises over San Diego harbor. [6.9.01]

6.9
Sails for San Diego

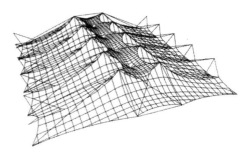

Computer generated general perspective illustrates the roof structure of the San Diego Convention Center, including the rain cover over the ventilation openings. [6.9.02]

Sails are efficient instruments which harness the power of the invisible wind to move the ship forward. Sails and their rigging are tensile structures; and no one understands their nature better than a sailor. Sailors have perfected the form of his sails for thousands of years to obtain a maximum of forward energy with a minimum of sail area, so that they can control their boats with ease and precision. They feel the effect of the wind by the pull on the sheet, and the reaction of the vessel by the seat of the pants. A good sailor knows a good sail by the beauty of its shape.

For the maritime city of San Diego, therefore, a fabric structure was an obvious choice to cover its new convention center. The possibility was discussed even before selecting a designer for the building. The four year old Nautilus Pavilion at Seaworld, though quite small, was the city's single example of a permanent fabric tensile structure roof. It had been an important precedent. As its designer, I had to appear before the city's Board of Standards and Appeals before receiving approval from the Building Department and the Fire Marshall.

Architect Arthur Ericson's firm was the principal designer in the team of architects who were selected for the convention center design. Ericson's headquarters were in Vancouver, another maritime city, where the fabric roof structure for Canada Place had just been com-

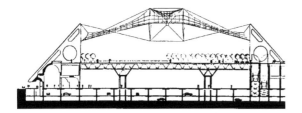

The typical cross section of the building shows 9from the top0: the fabric roof; the outdoor exhibit space; the main exhibition hall; and two levels of parking.
[6.9.03]

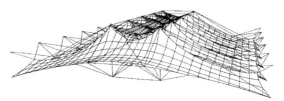

The computer generated perspective illustrates the network used for shaping and analyzing the roof. Bill Spillers was instrumental in programming the computer simulation.
[6.9.04]

pleted. Indeed, the most prominent features of his San Diego convention center design were the fabric roof of the outdoor exhibit area and the powerful concrete buttresses which support it. Between them they dominate the exterior shape of this landmark building.

The roof is supported by pairs of flying masts riding on long suspension cables which carry the loads to the two long sides of the building. Here they are anchored to massive triangular concrete buttresses which are spaced along the full length of the building, not unlike the flying buttresses of Gothic cathedrals. (Less than half of them support the roof encouraging the frequent question: "When are they going to built the rest of the roof?"). Framed into the buttresses are stunning glass-enclosed corridors, main elements of the building's circulation system.

The architectural concept of the building was well established when Horst Berger Partners was asked to join the design team as design and engineering consultants for the fabric roof structure. The conceptual design included the principal idea of the roof form: tent shapes are supported by flying masts which ride on main suspension cables; these, in turn, carry the load to the buttresses on two sides only, leaving the other two sides open and unobstructed by support elements.

But the design could not be built in this configuration. There was no structure to keep the free ends open. The rigid architectural model had not revealed this fact to its designers. A stretch fabric model would have demonstrated the need for a mechanism to resist the horizontal forces caused be the tension in the fabric and its supporting cables. Yet the open ends were a critical feature of the architects' design. A solution had to be found to make this part of the design possible.

We investigated various roof configurations with the help of stretch fabric study models, illustrating just a few bays and presented the results to the architects. Of the final two alternatives, comparing rectangular fabric panels with

Sails for San Diego

triangular ones, the former was agreed upon, though it deviated most from the original architectural design concept. As is always the case with tensile structures, the simpler, more flexible form was the more efficient structural form, and its greater simplicity made it the more attractive solution. The flow of the structural forces became clearly visible and its aesthetic appeal derived from the crisp delineation of its structural components and the sweep of its bright membrane surfaces.

In order to resolve the tensile forces acting at right angles to the main span we investigated various end support solutions and found none of them acceptable. We solved the problem by introducing a horizontal strut floating in the center of the roof which held the two ends apart. With its fork-like tips the strut pushes out the corners of the edge catenaries of the fabric membrane and keeps them in place, resolving the internal horizontal forces of the system in perfect equilibrium. Occupying the central axis of the roof structure, the horizontal strut was ideally located to serve a number of other functions such as the support of lights, loud speakers and sprinklers.

The roof consists of six sections spanning the full 300 ft. (91.5 m) between two rows of supporting concrete buttresses. Each interior section is 60 ft. (18.3 m) wide, matching the spacing of the buttresses but having splice lines halfway between two pairs of buttresses. The scalloped edges of the two end sections extend 30 ft beyond the plane of the last pair of buttresses. Large ven-

*Facing page, below:
One of several stretch fabric model, used to study alternative roof configurations. [6.9.05]*

The presentation model of the final design. It was one of many models built by Alexander Ngai for Horst Berger Partners. [6.9.06]

The roof, seen from the Mariott tower which abuts the convention center, is strikingly similar to the fabric model. [6.9.07]

tilation openings in the center of each section let warm air escape. To protect the space against rain, a secondary fabric structure, or rain fly, is suspended above the main roof, covering the ventilation openings. The delicate scale of its inverted tent shapes complements the bolder forms of the main roof, and strengthens the image of of a lofty vessel, sailing weightlessly through space.

The construction of each segment is quite simple: the flying masts and the main cables which support them are placed on a line midway between two buttresses. To reach up and over to their support points at the peaks of the fin-shaped buttresses, the twin main cables penetrate the fabric and split into a Y-shape. The ridge cables are located vertically above the main support cables. They attach to the flying masts near their top end, and flair out towards the edge of the building. Here they are connected to two sets of V-shaped edge support cables, which

The paired main support cables penetrate the roof surface at the splice line between two membrane panels [6.9.08], and split into a Y-shape to rise up to the buttress peaks.[6.9.09]

Looking at he completed roof, all structural elements are clearly visible: Suspension cables; flying masts; paired ridge cables; valley cables riding on the roof; the forked horizontal truss; ventilation openings; and edge catenaries. [6.9.13]

Sails for San Diego

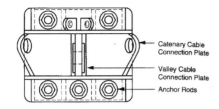

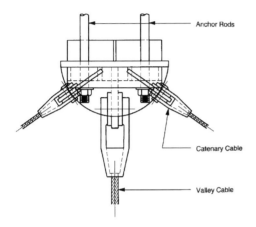

These drawings show the lower connection to the buttress. [6.9.10]

*The photo below shows the upper connection of the cables to the buttresses. [6.9.11]
The photo to its right shows the lacing strip which hold the valley cable in position. [6.9.12]*

are also anchored to the buttresses. This set of cables and masts carries the downward loads of the roof, and holds the membrane up.

Valley cables, placed in line with the but-tresses and connected to them 17 ft. above the floor, bow upwards in the form of an arch. They resist the upward loads (from wind suction and prestress) and hold the roof membrane down.

Edge catenaries on the end buttresses form small drapes, spanning 30 ft in plan view between the ends of the ridge and valley cables. The horizontal projection of the edge catenaries on the two open sides mirrors the vertical profile of the ridge cables. They swing across the face of the building, from one edge buttress to its opposite, passing over the two projecting support points formed by the forked horizontal truss. A grid of light cables hanging from the top of the flying masts and restrained against the bottom end of these same masts keeps this truss-shaped horizontal strut in position.

The main membrane has large openings in the middle of each bay. The rain fly is suspended from the tops of the flying masts and held down by anchorages connected to the center of the valley cables.

The cable system was erected first, independent of the membrane. The fabric panels span from ridge cable to ridge cable and from the buttress edge to the center line of the building. They we assembled on the ground, lifted in position, and connected to the ridge cables on either side. Valley cables were placed on a wear strip in the middle of each segment, and held in place with intermittent lacing strips.

The stressing of the main system was achieved by jacks near the top of the flying masts, pushing upward against movable collars which form the peaks of the main roof tents. The rain fly membrane was stressed by

The flying mast construction not only keeps the space free of columns but allows a 300 ft (91 m) wide open end which adds to the exhilarating sense of lightness. [6.9.14]

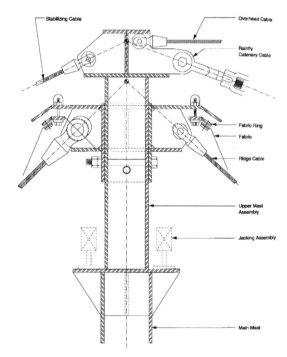

At the top of the flying mast, the main roof membrane is attached to a movable collar for stressing with hydraulic jacks. The rain-fly, attached to a fixed collar above, is stressed by pulling down towards the valley cables. [6.9.15]

jacking down against the center of the main roof's valley cables.

The building was completed in January of 1990. In a way, its progress defines a six year phase of my career. In December of 1983, my fifteen year old partnership with David Geiger had come to an end. Geiger Berger Associates had run its course. I had started a new firm under the name Horst Berger Partners with a few of my former associates. Chris Anastos, my longtime principal associate, had become my partner. We started out again with all the expected difficulties and anxieties. But we also experienced the clarifying miracle that a new beginning brings. The San Diego Convention Center project was the first major structure to be designed by the new firm, and its most important. It was part of that miracle.

In the spring of 1984, I had been asked to come to San Diego and give a dinner talk - to which architects and builders were invited - on fabric structures. In the autumn of that year, Jack Martin, president of John A. Martin Associates, the large west coast structural engineering firm, asked me to be the roof consultant on the convention center project. Budget and political problems had delayed the execution of the project for several years. It was not completed until January of 1990, just a month after Horst Berger Partners had closed its doors. (For continuity, I joined

Sails for San Diego

The San Diego Convention Center roof glows in the night. [6.9.16]

Severud Associates for the next few years, a period marked by the design and construction of Denver Airport.)

A major factor in my younger partner's decision to end the partnership was the difficulty of obtaining liability insurance. This was based on a false perception of the vulnerability of fabric structures - a perception which has since been corrected. In reality, our record was excellent, and insurance companies made good money on my projects.

The San Diego convention center roof remains one of the most elegant and expressive of my structures. It is the result of true team work. The citizens of San Diego love it. When the mayor tried to eliminate the roof in an effort to reduce costs, she was faced with heavy public opposition, which led to an open hearing in which she had to reinstate it. Though it may not be as dramatic a structure, to the citizens of San Diego it is nevertheless their "Sydney Opera House".

*The fabric roof of the new Denver International Airport shines in the night.
[6.10.01]*

6.10
Bringing Light to the Airport

About halfway along the thirty minute drive from downtown Denver to the new airport, the jagged outline of its huge terminal roof comes into view. Rising out of the vast plains, the high peaks and deep valleys of the 1000 ft. long tent structure echo the shapes of the powerful mountains in the distance behind. Although clearly an image of the dawning computer age, its forms are akin to the ancient Indian tipis which used to occupy these plains. The significance of its fabric structure is neither its size nor the novelty of its structural concept, but its integration into the architectural purpose and expression of the terminal building and new airport, which serves a critical public function 24 hours every day of the year.

The design returns to the sensible use of the most basic tensile structure form: the tent. Two rows of masts march down the length of the terminal's great hall, supporting a long progression of roof units to cover the 900 ft length of the space, and provide protective overhangs for the expansive glass walls which surround it. Ridge cables draped over the masts alternate with upward bowing valley cables. These cables are all that is needed to gather the snow and wind loads from the fabric membrane and transmit them to mast tops and roof anchors. No roof structure could be simpler, lighter, more direct in its purposeful form.

The space inside is defined by the gently disciplined

variation of its arched roof valleys, the soldierly order of the hard upright masts, and the rich play of light from the crisp triangles of the clerestory windows and the undulating translucent roof. This gives the space its special blend of serenity and exuberance, bringing to mind the mediaeval cathedrals with which this space shares the strict order of the ancient basilica.

The terminal's linear configuration is part of the simple layout of this airport, which is logical and easy to comprehend by the harried traveller. A spine, formed over an underground rail line, connects the terminal building at the south end to the row of concourses that straddle the line to the north. Runways are arranged to the east and west, parallel to the spine. Located in a large open plain, every component of the facility can be easily extended without interrupting the function of the busy airport. Additional terminal building space can be generated by extending the train track southward and adding another terminal building over It.

The idea of using a fabric roof for the terminal building came late in the design process. When Jim Bradburn, partner of the Denver architectural firm of W.C.Fentress, J.H. Bradburn & Associates, invited me in May of 1990 to come and explain the potential of fabric structure design and technology, his firm had just begun the working drawing phase of the design process. The conceptual and preliminary design had been assigned to Perez & Associates of New Orleans, and their work was complete. Fentress Bradburn's first task was to review the Perez design, and in the process find $40 million in savings to bring the project within budget. Having seen the just completed San Diego Convention Center, their hope was that a fabric roof would help not only with the budget and schedule problems, but also with the visual identity of the airport that was missing in the present design.

Curt Fentress had been looking for ways of relating the form of the building to the mountains which dominate the Denver landscape. He had investigated various structural materials and systems without satisfactory results. Tent shapes appeared to offer a natural answer. Of the carousel full of slides which I presented, it was the Riyadh Stadium roof with its huge peaks and deep folds which appealed to him most. And a slide of a Plains Indian tipi caught his attention. These were the images he had been looking for.

Together, we reviewed the existing design drawings and discussed the potential implications of the fabric roof on substructure and foundations, on building costs and construction process, on space function and energy consumption. I boarded the plane back to New York, taking with me plans and sections of the existing design and leaving behind an excited and expectant design team. Half of the 45 day review period was gone. There was little time either to develop a roof design which would do all that was expected of it, or to make a case for its acceptance to the owner - a case for a most unconventional structure in a town that was furiously divided over the need or desirability of a

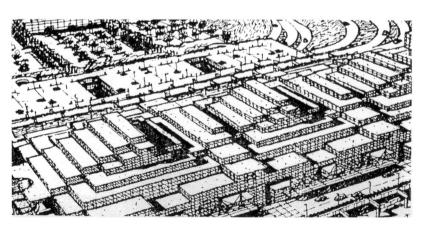

Perspective Sketch of the Perez roof design.
[6.10.02]

Bringing Light to the Airport

The fabric roof of the new Jeppesen Terminal, surrounded by garage buildings and runways' during construction.
In the distance are the Rocky Mountains.
[6.10.03]

new airport. During the fifteen minute drive to the old airport, the taxi driver had left no doubt as to the depth of the opposition: We don't need a new airport, he said. This is a waste of taxpayer's money and it will ruin the Denver we know. The usual state of confusion which I encountered inside the old Stapleton airport helped me to confirm that he was wrong.

To simply adapt the San Diego design to this project, as the architects had suggested, was clearly not the right thing to do. As I studied the plans for the terminal and sketched in overlays of various possibilities, the answer came into focus. The roof of the Perez design consisted of terraced levels of flat roof panels framed with steel trusses. It had two rows of internal columns which left a 150 ft (46m) wide space down the middle of the 900 ft long building. Tent masts placed in these locations would support ridge cables shaped to form the upper boundary of a series of tent shapes. They would be anchored to the framing of the low roof surfaces on the long sides of the building. Spaced 60 ft apart, they would provide enough room for a sufficiently deep fold so the fabric could carry the snow load safely. The 60 ft spacing of the mast supports would coincide with the proposed column grid of the building floors below, for which a foundation contract was already being negotiated. (Site preparation and excavation had begun months earlier in order to make possible an opening scheduled for Fall of 1993.)

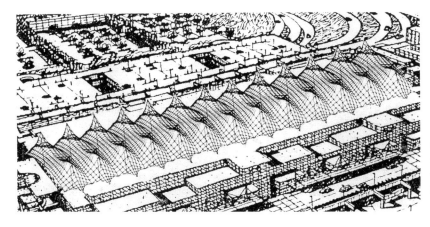

Overlay of isometric computer drawing showing the new fabric roof concept.
[6.10.04]

A few days after my return from Denver I faxed sketches and mailed a small study model which I had made. Mark Schlogel, with six years of computer modeling behind him, generated a first roof shape within a day and produced an isometric drawing which I pasted into a copy of an existing rendering. Here was the first image.

The circumstances under which I worked on this project were quite unusual and very difficult. Four months earlier Horst Berger Partners had closed its doors. To assure continuity, I joined Severud Associates, the structural engineering firm I had left 23 years earlier. A strange coincidence made this move easier: our offices were in the same building on 485 Fifth Avenue, near the corner of 42nd Street. After 23 years of running my own business, I was now an employee, even though my title was principal. By habit, I acted as a partner on my own projects. Mark Schlogel was among the people from Horst Berger Partners who had come to Severud with me. I had also brought our computer, loaded with the programs which we had developed over the years and tested on many projects. This made it possible to act fast, which was necessary.

Within a week, I returned to Denver for the first of a long series of design meetings, during which the project took on its final shape. Although the idea of using a fabric roof was attacked from many sides, the principal configuration of the structure was adopted by the architects, and accepted by the owners within just a few weeks. Perez Associates submitted a long list to the Denver City Council, detailing the reasons why such a structure would not work. Editorials in the local press, by pundits who hated the idea of a new airport to begin with, saw the fabric roof as the final proof of the irresponsibility and futility of the project.

There were a number of features which were changed or refined as the design progressed. The architects found it important to emphasize the entrances and the related internal three-part division of the terminal space by having pairs of tent units rise higher in two locations. The larger tent units were also designed to reach further out over the clerestory roofs. The result was that the repetitive nature of the roof geometry was largely lost. The transition from regular to unequal tents caused every unit in a quadrant to have a different shape. This made the design much more complex, and

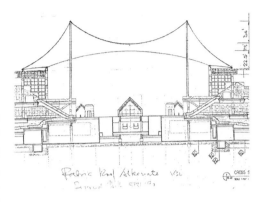

My first sketch of the proposed tent roof configuration.
[6.10.05]

The stretch fabric study model.
[6.10.06]

Opposite page: The construction photo shows the geometry and the details of the outer fabric.
[6.10.09]

Bringing Light to the Airport

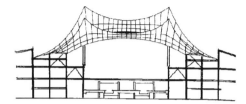

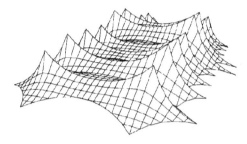

Incorporating the high units adjacent to the entrances, shown in section [6.10.07] and in computer generated perspective [6.10.08]

increased the cost of the roof by a large margin. Even today, I am not sure that this change was an improvement actually worth the effort and expense.

There was also a great desire to place glass elements in the membrane, in spite of the high translucency of the roof and the huge surrounding glass walls. The idea of placing strips of glass along the ridge was eventually abandoned after we tediously established the difficulty, if not the downright impossibility, of embedding a large rigid element within a soft flexible structure; instead, the covers at the mast tops were all to be glass, an idea which, in the final design, was only retained for the higher masts.

These design decisions had an impact on the structural configuration. The introduction of the higher units and the need to maintain a reasonably uniform stress pattern led to changes in the cable and membrane configuration which, especially in the areas adjacent to the higher units, required patient geometric tuning to retain a structure that would still work under wind and snow loads and avoid "ponding" of rain or snowmelt water. Ponding is a condition which requires particular attention in soft membrane structures as they experience

large deformations under load. Dips in the fabric surface caused by snow must have at least one area along their periphery where the surface slopes continually downward, so that water will automatically drain out. As Marty Brown, the project manager for Birdair, discovered during the patterning design process, there was indeed a small area of membrane surface adjacent to the high units which required the addition of closely spaced small cables under the outer fabric to ensure that ponding could not occur.

The reaction was always the same when people first heard about the fabric roof at Denver Airport: "But what about all that snow?" "All that snow," of course, occurs in the mountains above Denver. But Denver does experience heavy snow falls, and the roof is designed to carry them. The dead load of the structure is only 2 lbs/sq.ft. (0.1 kN/q.m), approximately 1/15 of a steel framed roof system, or 1/40 of a concrete roof structure spanning the same space. The proper evaluation of live loads is therefore critical, especially since the shape of the roof has a considerable influence on the magnitude and distribution of both wind and snow loads. For the study of both these load conditions, we engaged RWDI, a firm of special consultants in the area of environmental loads. RWDI's Dr. Peter Irvin had been our consultant on many structures, and we had come to value his contribution. Under Peter Irvin's supervision, RWDI conducted wind tunnel tests as a basis for our wind load assumptions. His firm also undertook a computer simulation of the winter weather over a 25 year period, to see what snow conditions the roof would have seen during that time. This became the basis for evaluating the snowfall for which the roof had to be designed. Snow accumulation due to drifting was studied in model tests in which the airstream was represented by flowing water, and snowfall by sand, a method RWDI has used successfully for many years. Several design load cases were assembled from the results of these studies which, again, became the load input for the roof design. In some parts of the valleys, this led to loads as high as 80 lbs/sq.ft., a load level which represents approximately 8 feet of snow.

The analysis showed that the fabric was capable of carrying the design snow load without assistance. Nevertheless, a set of 5/8" diameter cables, spaced approximately 40 ft. apart and running the length of the

Large steel rings were required on top of the masts to accommodate mechanical equipment and the related big covers.
[6.10.10]

Rings and covers had to be erected before installing the fabric. Here, an end panel of the outer fabric is in place, attached to the diagonal panels. [6.10.11]

Bringing Light to the Airport

Snow drifts were investigated with the help of model tests. In this simulation by RWDI, water replaces air and sand replaces snow. [6.10.12]

Half of the outer fabric is installed. In the foreground are half sections of mast top covers. The crane at the right is erecting a window frame. The redundancy cables in the roof surface and their rip stop connector strips are visible. The "worker" facing away from the camera is my wife on her first trip to a construction site. [6.10.13]

building in sections attached to the ridge cables and valley cables, was placed just under the membrane. Their primary purpose is to create redundancy: should a panel of fabric have to be replaced or should a rip occur, they will take over the structural function of the fabric, and thereby guarantee the integrity of the structural system. They also reinforce the fabric, and effectively reduce fabric stresses and deformations. They are part of a rip-stop system which would contain any accidental rip within the 40 ft length between any two such surface cables. To achieve this, fabric strips which are heat-sealed to the outer fabric contain the cables in a staggered loop pattern similar to a door hinge.

There is only one other set of cables in the membrane surface: long diagonals reaching from the mast tops down to the tie-down anchors just outside the clerestory walls. They allow the geometric separation of inner panel and end panels, which is necessary to make fabrication and erection practical. They are also shaped to help resist the wind suction on the end panels, which are not unlike sailing boat spinnakers.

At the tie-down anchors, these diagonal cables meet the valley cables, edge catenaries and outward and downward anchor cables. To bring all these cables together in one connector was not a simple task. And to do it with one typical connector which would fit all the

178

varying geometric conditions along the periphery of the roof was particularly difficult. We called Ed de Paula solution to the problem the "octopus". After Mark Schlogel had left, less than half way through the design, Ed, then one of Severed associates and now a partner in the firm, took over as project manager. He had no prior expert-ecce with fabric structures, but he learned fast and kept the project on track.

The introduction of large caps on the mast tops to cover mechanical venting equipment resulted in unusually large tension rings to terminate the top ends of cables and membranes. These elements had to be erected before the membrane went into place, making it impossible to stress the structure from the top. Ed and I studied the possibility of jacking at the bottom of the masts, which had worked well on several earlier projects. But here, the forces were too great, and the details became unruly. Searching for a way out, it occurred to me that there was one other location which was, indeed, more suitable for stressing the membrane structure, and that was the octopus connectors. Using hydraulic jacks to push the octopus connectors downward would cause the valley cables to pull down on the fabric, and thereby put load into the ridge cables and all the other components of the system. A preliminary study of the roof geometry before and after stressing indicat-

Valley cables, edge catenaries, vertical and lateral anchor come together in one "octopus" connector. It is shown (above) before stressing; [6.10.14],

and (below) in the completed structure. [6.10.15]. The drawings (below left) show both the "octopus" assembly and the anchor to the building structure below. [6.10.16] [6.10.17] (below left)

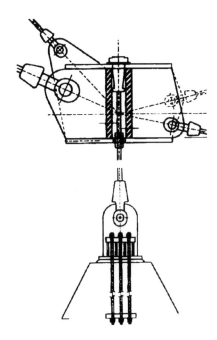

Bringing Light to the Airport

The inflated tubes which form the closure panels between roof membrane and window wall frames are installed. [6.10.18]

The inner liner is in place, hiding the closure tubes. [6.10.19]

ed this to indeed be a feasible alternative. Because of the easy access it was, actually, the simplest stressing process conceivable.

Although the roof membrane has a total sur-face area of 377,000 sq.ft. (35,000 sq.m), no expansion joints are required. This is due to the flexible nature of the folded fabric structure system, which easily absorbs temperature deformations with negligible change of stress in its structural compo-nests. Inside the perimeter walls, the roof consists of two layers of fabric formed by the outer membrane described above, and an inner liner located approximately twenty four inches below the outer membrane. The inner liner's purpose is to provide thermal insulation and acoustic absorption. It is supported by its own system of ridge and valley cables; around the periphery it is connected to the upper members of the window wall framing system.

The fabric roof itself undergoes large deformations under both wind and snow load conditions. This is typical for a tensile membrane structure which absorbs loads by change of shape as much as by increase in stress. In order to cover the gap between the rigid walls and the flexible membrane roof, we had to develop closure panels which could accommodate the roof movements without causing unacceptable stresses in the fabric or the walls, and yet possess sufficient strength to resist the wind loads. The final answer to this difficult engineering problem was to introduce inflated fabric cylinders of up to 3 ft. diameter over the clerestories, and up to 5 ft. diameter over the north and south walls. Interconnecting them by means of plastic hose sections, while providing them with spring operated valves and maintaining their inflation by means of pressure from small pumps, gave them the required flexibility. Because they are positioned just above the inner liner surface, from the inside they are not visible, while on the outside they are tucked deep under the overhangs.

All glass walls are framed by cable-supported structures; the peaks of the clerestory windows are held in position laterally by stay cables; the vertical mullion are reinforced by cable stays similar to the mast stays on

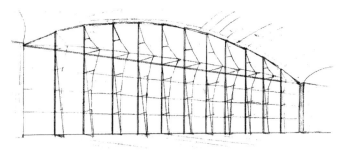

sailing yachts.

The support system of the huge glass area of the south and nor-th walls is more complex and dramatic. These walls are large, free-standing structures. The south wall is 60 ft. high in its center, and spans 220 ft. horizontally. The outer fabric roof floats right over it, while the inner liner is attached to the top and pulls inward, adding to the horizontal wind pressure acting on the glass surface. The wall structure consists of a two-way cable-reinforced strut system, in which the vertical members are the principal load carrying components. They are spaced 15 ft. apart, and have a common shape with varying dimensions. Two main cables, in the form of an A-frame, support the main node point 37 ft. above the floor, where a horizontal strut is reinforced by two bow-string cables, one inside and one outside, spanning the whole width of the building. The lateral struts which generate the bowstring curve also form support points for the cables of the vertical structural elements, to which the window wall is attached with short ties.

Although complicated to describe, the manner in which this cable and strut network carries the load is easily discernible. In fact, when I first sketched it, all the pieces fell into place instantly. And as in the flying buttresses of Gothic cathedrals, when forces are resisted by visually expressive systems, structural form becomes natural, meaningful, and pleasant to the eye. Although the support system of the south wall is light and elegant, the deflection at the top of the wall under maximum load conditions is only 3 in.

Construction of the terminal enclosure was a linear process starting at the north end and proceeding towards the south. The masts were installed several bays ahead of

This drawing was my quick response to the suggestion that the rigid framing planned for the south wall be replaced by a tensile structure. The two-way system is supported by three sets of cables: A-frame shaped stays support a continuous horizontal member spanning the full 220 ft. Cables draped in the vertical plane complete the structure, stiffening the vertical mullions of the glass wall. [6.10.20]

The printout from a preliminary computation shows how the structure performs under load. Total deflection at the top under maximum load is 3 in. [6.10.21] The section (across) shows the actual geometry at the center mullion. [6.10.22]

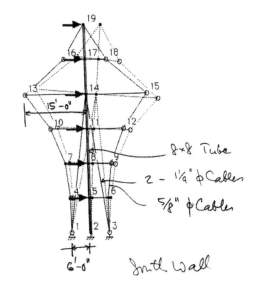

Bringing Light to the Airport

The completed south wall.
[6.10.23]

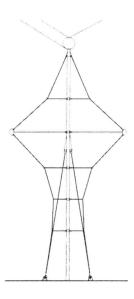

fabric installation. The top rings and mast top covers were then assembled at the bottom of the masts and hoisted up with winches. The fabric was installed one bay at a time, starting with the outer panels and followed by the much larger inner panels. In both cases, the adjacent fabric panels were clamped to the ridge cables on the ground, and were then raised with the ridge cables. The installation of the outer panels was completed by spreading and attaching the edge catenaries, and clamping the fabric to the diagonal cables. Similarly, the inner panels were lifted with the ridge cables, and then spread and attached to the valley cables and diagonals.

The addition of each new unit added stress to those already installed, bringing them closer to the final shape. After all units had been installed, final stressing was achieved by jacking down on the octopus connectors in a number of passes, until the design stress level was reached.

This was the most exciting construction site I have visited in my long career. The level of activity was hard to imagine: roughly $100 million worth of construction work was completed every month. Driving through the site, I had to be alert to stay out of the way of huge 100-ton pieces of earth-moving equipment hurtling through

Construction started from the north, moving towards the south. Masts were erected first. Then top ring and cover assemblies were lifted into place. Valley cables were strung across, and the fabric installed a bay at a time. [6.10.24]

the land-scape at 50 miles/hour.

The most vividly remembered of my many trips to the site was right after Thanksgiving in 1992. All units but the last had been installed, but no stressing had taken place. As I got off the plane the evening before the site visit, I had noticed skiers picking up their gear. Outside, snow was on the ground; in fact, as much as a foot had fallen. Next morning at the site, the snow accumulation was quite visible from beneath the single membrane. Some of the units had enough sag for the valley cables to show slight bagging in the middle. Ice had formed, since the sun had made it warmer on the outside than below the roof. Considering the lack of prestress, the roof behaved excellently.

The best part was that everyone in Denver interested in or related to the airport project had trouped out the day after the snowfall to see the roof. They had stood under it, and seen how the snow made dark shadows in the translucent fabric. It looked perfectly safe - and it was perfectly safe. They went away feeling much more confident that the ephemeral structure, which they had accepted with some doubt, was indeed going to work.

The roof was completed in the spring of 1993, ahead of the construction deadline. The curbside canopies on both sides of the Jeppesen Terminal were installed later; they cover the sidewalks, and cantilever over the roadway for the full 1000 ft. length. Their supports are circu-

For transportation, fabric panels were rolled around a tube which also assists in moving them into place. [6.10.25]

Bringing Light to the Airport

A foot of snow falling during construction demonstrated the capacity of the roof to carry snow. The drift patterns were very similar to those in the snow test model. [6.10.26]

lar piers spaced 30 ft. apart, although at the main entrances their spacing widens to 60 ft. A set of horizontal and vertical struts on top of the piers holds the fabric membrane upwards and outwards. Tie-down points in mid-bay between the piers hold the structure down. The tie-down points also provide drainage for rain and melting snow. This function, together with the need to avoid large snow slides being dumped into the roadway, determined the shape of the structure; clearances over and around components of the terminal building put further limitations on its geometry.

The site under snow, taken on the same day. [6.10.27]

Light Structures
Structures of Light

The interior photograph of the terminal, taken in 2002, shows the space in use on a regular day. [6.10.28]

The curbside canopies extend the full length of the building on both sides, and cantilever over the sidewalks and part of the road. Drains are combined with tie-down cables. [6.10.29]

Bringing Light to the Airport

The great tent in the fields.[6.10.31]

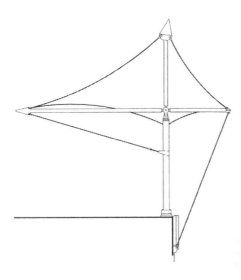

The drawing delineates the canopy structure. [6.10.30]

Problems with the electronic baggage conveying system prevented the airport from opening on time, and gave detractors in Denver and the media throughout the country something to gripe about. (I would venture to predict that the new baggage system, once the bugs are out of it, will become the new toy which every airport will want to have.)

A picture of the roof appeared alongside all the bad news about the baggage, and made this fabric structure one of the the best known buildings in the country and the world.

Looking up at the 300 ft clear span roof of the San Diego Convention Center, all structural components are clearly visible: suspension cables supporting the flying masts; paired ridge cables; shadows of valley cables riding on the roof; the forked horizontal truss; ventilation openings and edge catenaries. [7.01]

7 Covering very large Spaces

Size has an important qualitative impact on structures. An elephant is not a large ant as you can see by comparing the legs. The weight of structures grows with the third dimension. Large structures need a totally different approach to keep them economical. [7.02]

"I have a tiger by the tail," David Geiger told me one day in the summer of 1968. The tiger was the design of the enclosure structure for the United States Pavilion at the 1970 World's Fair, in Osaka, Japan. David was the structural consultant on the design team which had won the competition for the project, and he needed my help. I was interested. The conversation led to the formation of Geiger Berger Associates. Large span fabric structures became the main activity of the firm for the 15 years of its life.

A few years earlier, before leaving to earn his doctorate at Columbia University, David had worked on the team I headed at Severud Associates. Research for a paper on long-span steel structures had led him to a Russian engineer's proposal for a low profile air-supported membrane roof. David developed his own concept and proposed it for the Osaka project. It was a design concept that was to dominate large span roof construction for more than a decade.

Enclosing large spectator sports facilities had become one of structural engineering's great challenges. The Astrodome in Houston, Texas, the first full size domed stadium, was completed in 1968, just two years before the Osaka World's Fair. Its radial steel dome covered a circular space of 658 ft (200m) diameter, stretching the capacity of rigid structural systems to its limits.

Although it was a tremendous achievement, as with earlier dome structures it suffered the disadvantage of a difficult construction process that called for temporary construction supports, since arched structures function only after the last section, the "keystone", is in place. The introduction of cable-supported roof systems more than a decade earlier had opened the way to easy construction. The 450 ft (137 m) span roof of the Madison Square Garden arena in New York, designed by Severud Associates in the early 1960s, uses cables in their simplest form. Its radial cable system is a reversed dome, hanging from a circular ring beam.

While this configuration works well for the multi-tier Madison Square Garden arena, cables draping downward are often not acceptable for stadium roofs where - as in baseball - the center of the roof is ideally 200 ft (61 m) above the playing field, while the highest row of seats might rise no more than 100 ft. (30.5 m).

The new air-supported roof system was a natural solution to this geometric problem. It combined the easy installation of a hanging cable system with the desired form of a dome: low along the periphery and high in the

Left: The Astrodome in Houston, Texas, completed in 1968, was the first full size covered stadium. A trussed steel dome of 658 ft (200 m) diameter supports translucent rigid roof panels. [7.03]
Above: The radial cable system for the Madison Square Garden arena in New York city is much lighter and much easier to construct. [7.04]

The low profile air-supported roof of the US Pavilion in Osaka illustrates David Geiger's ingenious geometry: a diagonal cable net contained by a super-elliptic ring beam. [7.05]
The entire 100,000 sq.ft. membrane arrived on one truck. [7.06]

Covering Very Large Spaces

The air-supported roof of the Pontiac Stadium, near Detroit, MI, spans 722 ft (220 m) by 552 ft (160 m). It took weeks to erect and its phenomenal low cost made enclosed stadiums affordable. [7.07]

center. It did so with the help of internal air pressure, which inverted the shape of the tensile cable roof. By balancing the forces of a two-layer diagonal cable net with the super-elliptic shape of its ring beam, David's ingenious geometry achieved this with a mini-mum of material.

In Osaka, the entire fabric roof arrived at the building site on one truck. It was unfolded on a huge platform on which the cable net had been laid out beforehand. After lacing the fabric to the cables, the space below was pressurized, lifting and stressing the roof and generating a domed space of great elegance, filled with diffuse daylight. It was clearly the simplest, fastest, and most inexpensive method of covering a large space.

The success of Osaka became the foundation of a campaign which offered low profile air-supported structures as the solution for enclosed stadium roof installations. David devoted himself to the concept with tireless energy and enthusiasm.

He replaced PCV-coated fabrics with Teflon-coated fiberglass, a material mainly produced by Chemfab of Bennington, Vermont, for conveyor belts in industries with chemically aggressive environments. This made the roof structure non-combustible and dirt resistant, while increasing translucency and offering a life span of 20 years or more. In the wake of Osaka, Geiger Berger Associates designed seven large air-supported domes, which were built between 1975 and 1983, and which were all fabricated and erected by Birdair Inc. of Buffalo, N.Y. (In 1985, Takenaka added the Tokyo Dome under a license agreement with Geiger Berger.) Among the first was the Pontiac Silverdome, a 78,000 seat football stadium for the Detroit Tigers. It took two years to build. Two stadiums with rigid roofs - the steel-framed Superdome in New Orleans and the concrete-framed Kingdome in Seattle - had started construction earlier, and were completed later at considerably higher cost than an air-supported roof. Although the air-supported roofs of these domed stadiums have served well, there is a continuing problem relating to the need to pressurize the entire building. Construction economy requires maintenance of the pressure at a level below 10 lbs per sq. ft., even in locations where the design snow load is much higher. Excess snow load is avoided with the help of mechanical systems which redirect warm air to the roof surface in order to melt

The saddle shaped roof of the Raleigh Arena was the first cable supported large span roof. [7.08]

snow as it falls. But when the snowfall is too heavy and too wet to be melted, and especially when the system has not been set in advance to anticipate the storm, manual removal of the snow becomes necessary. The owners of stadiums with air-supported roofs have developed snow removal techniques that primarily involve emergency crews, who use warm water running in hoses. Yet a number of dramatic deflations have occurred over the years, and this has eroded confidence in the reliability of air-supported roof structures.

As a result, a number of owners are presently investigating ways of replacing their air-supported domes with static structural systems which are not dependent on the maintenance and operation of mechanical systems to keep the structure stable. Indeed, at the time of writing I am actively involved with Light Structures Design Consult-ants in the development of replacement designs for three of the seven facilities.

From the outset, I had worked on tensile structure solutions as alternatives to the air-supported roof system. These solutions included an arch-supported roof to cover an existing baseball stadium in Bayamon, Puerto Rico (illustrated in sec-tion 8), and an A-frame supported tent structure for an arena in San Cristobal, Venezuela. Although neither was built, the latter became the point of departure for the design of the Riyadh Stadium roof, which is described in Section 6.6.

A tensile form which makes use of draped cables in the most direct way is the saddle-shaped roof. Like air-supported roofs, it consists of two intersecting sets of parallel cables. But here, the cables curve downward in one direction, and upward in the other. Putting stress into the cables causes the two sets to press against each other, giving the roof stability and rigidity, and providing a clear path for carrying upward and downward loads. Engineer Fred Severud, working with the architect Nowicki, had first devised this concept for their design of the 1953 Raleigh Arena in North Carolina. During my eight years at Severud Associates I had worked on a number of saddle roof designs. Their only inefficient element was the supporting ring beams: snow or wind loads increase the forces in the load-carrying cables, and let the opposing set of cables go slack, thus imparting large bending moments to the ring beam, and requiring it to be large in width and heavily reinforced.

In the wake of the Osaka success, we were asked to be the engineers for the Capitol Bullets Arena, near Washington, D.C. A saddle-shaped cable roof offered an excellent solution for this facility. The concentration of spectator seating on the two long sides of the rectangular play area (for basketball or ice hockey) fitted perfectly inside a warped 400 ft (122 m) diameter circular ring which is high on two opposing sides. To eliminate bending of the ring we introduced two additional sets of stabilizing cables running straight across the building like the strings of a tennis racket. Because of the ring's warped

Covering Very Large Spaces

Saddle dome structure for the Capitol Bullets Arena in Landover, Md. Note the construction workers in a roof panel. Straight cables stabilize the thin ring beam. [7.09]

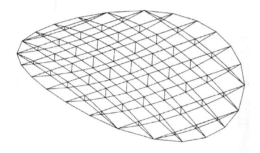

The double-layer cable net of my patented cable dome: load-carrying and restraining cables are combined with vertical separation struts to optimize the structural shape. [7.10]

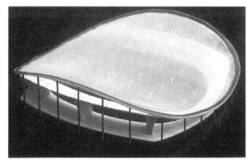

This model shows the proposal for a Kansas City football stadium with a saddle dome roof. There is a clear path for rain and snow. [7.11]

shape, one set is high and outside, the other low and inside. These cables restrain the ring, and center the forces. The resultant square cross section, although only 8 x 8 ft. (2.4 x 2.4 m), is nonetheless adequate. Because translucency was not required, the roof was covered with prefabricated metal panels, each filling one square formed by the intersecting support grid.

Later, I developed this concept further into a design which I named the "saddle dome". It retains the four sets of cables, but rearranges them in two separate nets which are held apart by vertical struts. Each net contains a set of load carrying cables and an intersecting set of restraining cables. The struts greatly increase the curvature of the load-carrying cables, permitting the use of smaller cable sizes. In 1985, the proposed roof of a football stadium in New York City demonstrated the efficiency of this concept. It became one of three patented structural concepts for which Horst Berger Partners licensed Obayashi Corporation for use in Japan.

Another was a structural system named the "cable dome". I developed it in the Spring of 1983 for the proposed domed stadium in St. Petersburg, Florida. Initially HNTB, the project's architects, had only wanted a shade structure over the seating of an open stadium. In meetings with the owners, however, this modest cover grew, step by step, until it became the roof of an enclosed, air-conditioned facility.

The principal idea of this tensile roof system was a tent structure, with poles that did not go to the ground, but were supported by suspension cables carrying the load to their end supports. The primary elements of this system are similar to those which support the roof of the San Diego Convention Center. I felt that this was a tensile structure solution which could compete with the air-supported roof because, despite its apparent complexity, it was simple.

In this original St. Petersburg design, both ends of the 24 identical support cables are anchored to and cut across a small segment of the circular compression ring. Each cable supports two struts. Two such cables intersect at the support point of every strut, so that each strut rests on two cables. The upper ends of the struts are tied back to the ring by restraining cables. These can be strictly radial, or can follow a shape similar to the support cables. The tops of the flying poles form the support points for the next tier of support cables, which follows the same geometric pattern. For the 650 ft. (198 m) diameter dome for St. Petersburg, three tiers were sufficient to frame the entire roof.

The design drawings were completed in June of 1983, but by the time working drawings were started in Spring of 1984, Geiger Berger Associates no longer existed. It had never been an easy partnership. The invention of this cable dome system became an important factor in the breakup of the firm. Because of the unparalleled success of David Geiger's air-supported domes, he had come to regard domed stadiums as his domain in the firm - a position which he defended ferociously. He could not except the idea that his dominance in this area would end. There was no longer room for both of us in the firm.

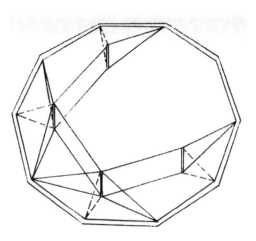

This drawing shows the principal cable and strut elements which make up the structure of my patented cable dome system [7.12]

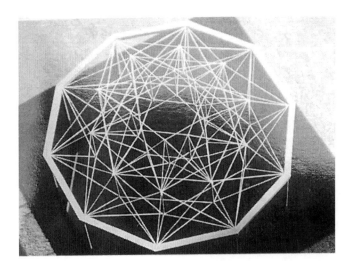

The model illustrates a two-tiered, 10-sided cable dome structure. [7.13]

Covering Very Large Spaces

David Geiger's St.Petersburg cable dome under construction, [7.15]

On the right is part of a design drawing of the original St. Petersburg stadium roof. [7.14]

In December of 1983, I left and formed Horst Berger Partners. Around that time, the St. Petersburg designers left HNTB and joined HOK, taking the project with them. In the subsequent re-alignment of highly competitive architectural and consulting firms competing for the design of large sports facilities, David Geiger ended up on their side, Horst Berger Partners on the other. It was the beginning of fierce competition.

David developed his own cable dome system. It has a radial configuration: the struts rest on concentric rings which are tied back by radial support cables. It has the most direct force flow and therefore requires the smallest amount of cable material. But it lacks the simplicity of construction and the redundancy which I consider critical. He had tried his system out on two arenas in Korea before using it in the St. Petersburg "Sundome" which, after various delays, was successfully completed in 1987.

It was at an Istanbul conference on domes that one of the presenters reminded me of the directness of the flow of forces in a classical dome structure. Being aware of the tortuous path of this flow in a cable dome structure, it occurred to me that adding arches to a cable net structure might lead to an excellent combination. Arches would transmit gravity loads to the ring supports in the

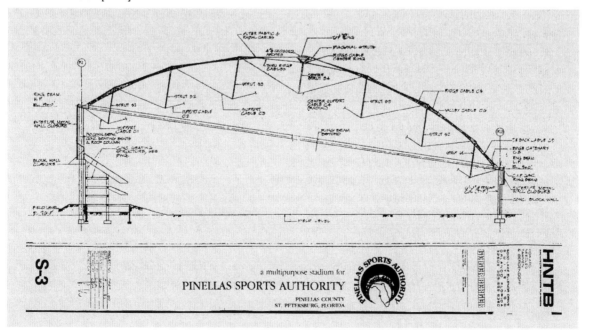

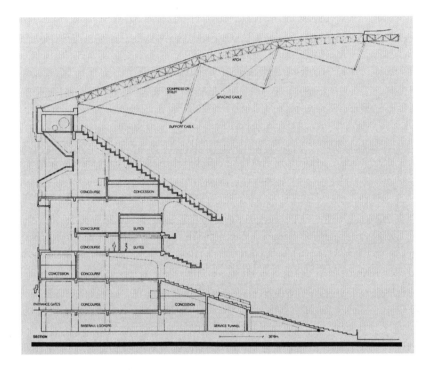

Cross section through design for a Tampa baseball stadium using my arch-reinforced cable dome structure. [7.16]

most direct way. The combination would not eliminate the visual interference caused by the downward slope of the support cables but it would greatly reduce it. This is because the much lighter cable dome components would serve mainly as a means of erecting the arches and keeping the arch system stable and more balanced.

I first proposed this combination of a cable dome with arches for the design of a proposed stadium in Tampa, Florida. At my presentation to the Georgia Dome architects in Atlanta, I carried a model illustrating this new structural system. But they were sold on a pure cable dome, a legacy of the persuasiveness of my late partner, David Geiger, who had been involved in preliminary studies for this building. Ironically, the emphasis on arches cost me the job. Weidlinger Associates were selected as engineers, and the dome was completed in 1993. Its configuration is not as clear and simple as either of the two systems previously described: its roof membrane consists of large four-point panels; those near the edge have a reversed curvature which not only gives the dome its strangely unfamiliar appearance, but also creates shallow triangular areas along the periphery which are difficult to drain.

After studying many structural systems to replace existing air-supported roofs, the deflation of the UniDome at the University of Northern Iowa during a

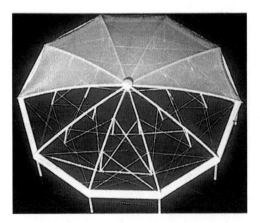

This model illustrates the integration of arches with the cable dome system, providing the most efficient path for the various types of load a roof has to carry. The arches give the roof the classical form of a dome. [7.17]

Covering Very Large Spaces

UniDome at the University of Northern Iowa: the original air-supported roof.

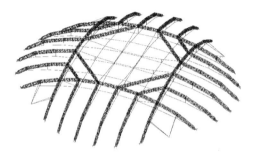

The structural system of the replacement roof is shown in the drawing above: an arch grid stressed to a cable grid below. [7.18]

The photo below it shows the arch grid under construction.[7.19]

At right: the completed new roof of the UniDome,the stainless steel opaque periphery and the translu- cent fabric roof in the center. 1998. [7.20]

snow storm in December of 1994 made it necessary to select a structural scheme for its replacement. A combination of arches and cables proved most efficient: light arches copy the plan geometry of the existing cable grid, rising a distance above the present roof line; the ring beam and the existing cable net are retained; and an internal cable ring reinforces and prestresses the ring beam. Vertical cables connect the arch grid to the cable net below, and the two are stressed against each other. In the center, fabric is draped over arches and pulled down by segmental valley cables. Around the periphery, solid highly insulated roof panels span between the arches. This creates optimal conditions for energy efficiency The construction cost is indeed more than that of an air roof, but the life cycle cost, which takes energy and operational expenses into consideration, is lower. And the fear of a deflation is gone.

This design was eventually selected and - with some adjustments - carried out. The ring beam is reinforced by posttensioning. The existing cable net from the air roof is re-installed and the arch grid is stressed against it. The structural steel required is only 11 lbs./sq.ft. The cost of the roof is only $60 for each square foot. The construction fit mainly into the spring of one semester and the summer. (Completion was missed by one month).

(Unfortunately the stainless steel standing seem roof hat leaks in some places. This was counter to our very good experience with a similar roof at the the University of Wisconsin's McClain practice facility. It convinced me that a fabric membrane, spanning between arches, was clearly the better answer. Not only are fabric membranes reliably waterproof, but the entire detailing of the steel becomes enormously simplified. And an opaque

Retractable roof design for the Milwaukee Brewers baseball stadium. The radial segments run on a circular track. 1986. [7.21]

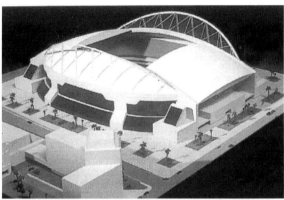

Retractable roof design for a multipurpose stadium in Phoenix, Arizona, (including football and baseball).The rectangular panels run on parallel tracks. [7.22]

roof can be very inexpensive).

The last decade has shown a renewed interest in enclosed stadiums, preferably with retractable roofs. In good weather, people prefer to watch sports events outdoors. But they like protection against rain or snow. Also, to be economical, stadiums need to serve a wide range of activities. A retractable roof can adjust to all conditions.

There are only two principal geometric configurations which permit a roof to retract: the circle and the rectangle. Movable roof elements can roll on a circular track or on two parallel straight tracks. There are no other solutions. Again, we are back to the basic dome forms: radial and orthogonal. In the 1980's I was given the opportunity to develop designs for both of these. Unfortunately, they were not built then.

The roof concept I proposed in 1986 for the Milwaukee Brewers stadium is an example of a radial system. This simple structure has only two movable sections. One scaffold enables the assembly of one pair of arches at a time, and their subsequent displacement to make room for the next pair.

The roof was designed to be part of a general renewal of the existing stadium. This never materialized. More than ten years later the Brewers decided to built a new stadium, which was completed in 2002. I is also of the

Covering Very Large Spaces 197

radial configuration consisting of five sections. The design uses conventional materials and systems. When it became known that the design was over budget, Birdair asked me (with DeNardis Associates) to look at an alternative design using fabric. It showed that considerable savings were possible. But the schedule was to far advanced to change design.

For the design of a retractable roof for a multipurpose stadium in Phoenix, Arizona, I proposed the parallel track approach to HNTB, the architects : the tracks were suspended from two large exterior arches. It was the winning design in a tough competition, but as so often, it was not built. This concept, for which I received a patent, won the competition. Implementation depended on a major league football team coming to Phoenix. When the St.Louis Cardinals moved to that city a few years later, the right moment seemed to have come. That is almost 20 years ago. A similar scheme (without the overhead arches) is being built, designed by HNTB with Peter Eisenman as architects and   as structural engineers.

Functionally, the better approach is a roof spanning the entire width of the stadium, and rolling on horizontal tracks. A number of stadiums have been covered with retractable roofs based on conventional steel trusses. A more elegant solution is a cable-truss. It is the lightest structure to achieve such a large span. Its rigid steel core

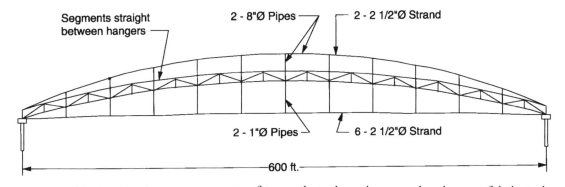

This drawing shows a proposed cable truss system to achieve a clear span for a stadium roof structure most efficiently. The core truss is shop fabricated in transportable sections. It doubles as construction platform and as catwalk. [7.23]

of trussed steel sections can be shop prefabricated and assembled on site. It can be used as a cat walk, serving lighting, scoreboards, and stage frame supports. The roof surface can be of fabric or rigid panels, with the fabric panels spanning between the trussed steel cores, or draped over the exterior cables. Structure, function and form once again form an integral unit.

The presentation model for the roof of Bayomon Stadium in Puerto Rico. The arch is made of soldered steel wire, the membrane of Spandex stretch fabric, and the cable net pattern is of flexible silver string. [8.01]

8

Physical and Mathematical Models

As the many examples in this book have illustrated, the form of a tensile surface structure is critical to its structural function. Its geometric order reflects the structural laws which keep the surface forces in equilibrium - a condition which cannot be satisfied by combinations of arbitrary geometric shapes. Finding the proper form, therefore, requires the use of design tools with which neither architects nor structural engineers are commonly familiar.

The surface form of a tensile structure can best be described as a graphic representation of a complex field of forces in equilibrium. A simple two-dimensional example is the drape of the main cable of a suspension bridge. This drape is the result of generating simultaneous equilibrium in all connection points or nodes, where the vertical suspension ropes connect to the main support cable.

The ultimate three-dimensional example is the movement of the planets in space. Each planet travels on a geodesic pathway determined by balancing the multitude of ever changing gravitational forces acting upon it. Being one of a billion bodies in the universe, a planet's behavior can only be understood as part of an interactive force system involving the entire universe, and even uni-

The shape of a suspension bridge is a simple, two-dimensional example of an equilibrium force line. It is generated by simultaneous balancing of the intersecting cable elements at each node point. [8.02]

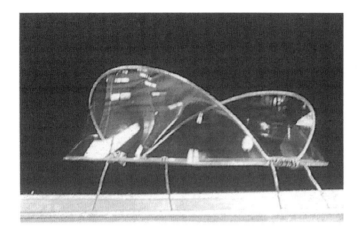

Modeling with uniform clear film surface. [8.03]

verses unknown to us. In its own modest way, a tensile surface is an interactive universe of forces. Mathematically, it is represented most conveniently by a net of intersecting force lines. When correctly shaped, the forces in all the lines intersecting at anyone node are in equilibrium with the weight of the structure acting on this node. Once the support conditions and the laws governing the force pattern are established, the form of a tensile structure will follow. The primary tools for exploring and defining the shape of tensile surface structures must therefore be three-dimensional. They include physical and digital models.

Modeling with elastic string material. [8.04]

Various physical modeling techniques have been explored. Anyone who has played with soap bubbles remembers their perfect spherical form. Blowing air into a closed soap film increases its internal air pressure, which is in turn resisted by surface tension. The spheres are an expression of perfect uniformity: their surface tension is uniform, at every point and in every direction.

Soap film itself is much too flimsy for modeling structural shapes, but clear films with similar characteristics, which harden into a rigid surface, have been used for modeling shapes with continuous edges. Networks of elastic strings are another means of modeling shapes which are not too complex. Nevertheless, by far the best modeling tool for studying and illustrating design concepts of tensile surface forms is stretch fabric. Models made with this material are most realistic in their structural behavior. They are relatively easy to make, and the end product is usually handsome enough to serve as a presentation tool. By involving the hands and eyes, this process of model making draws on powerful resources

Modeling with stretch fabric (Spandex Lycra # 13). [8.05]

Physical and Mathematical Models

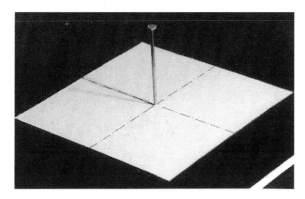

The base with the mast. [8.06]

Stretching and pinning the fabric. [8.07]

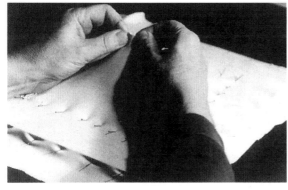

Adjusting the top of the mast. [8.08]

deep inside our minds, which I believe contain creative memories not reachable by conscious, analytical thinking.

A light and highly stretchable material, with similar elasticity in both directions, is best. Pantihose is good for small, exploratory models; but for a serious study or presentation model, I prefer a sturdy stretch fabric such as spandex, preferably with a shiny finish which is very effective with model photographs.

Over many years of model making I have developed a process which I will now demonstrate on a simple roof structure supported by a single mast. I will use a square form anchored directly to its base. By avoiding elevated support members, which require a strength sufficient to resist the stressed membrane, the model making process is greatly simplified.

I begin by cutting a piece of foam board with ample edge space around the structure, and then draw or paste on it a copy of the floor plan. I use a thin brass tube for the mast, with a pin at its base to keep it in position. A small button placed on top acts as support ring and prevents the mast from puncturing the fabric. A few dabs of glue keep the button from sliding off the mast.

I then cut a generous piece of fabric and drape it over the mast. I stretch the fabric by going around the periphery, pulling the fabric outward in gradual steps and attaching it to the foam-board with pins. Starting with just a few pins in the corners and along the sides, I repeat the process until the fabric is taut and held with pins not more than 2 or 3 cm apart. While stretching the fabric, the mast has to be straightened several times. I lift the fabric up near the top, and push the mast into position. As the stress increases, the fabric becomes more transparent, and the position of the mast becomes easily visible. (A small triangle can be held behind the mast to assure its vertical alignment.)

The structure is beginning to take shape. Now is the time to make any final adjustments to layout and mast height, so that slopes, curves and stresses are sufficiently uniform throughout the surface, and the structure is well proportioned. In preparation for installing the cables, I place pins in the exact anchor locations at the corners, and in the mid point of the edges. I also put one pin through the top of the mast to fix it. Internal cables are needed to transmit the load to the mast top, since a point support at this position would cause the fabric to be overstressed. From various alternatives, I decide on the use of four cables in the middle of each side, generating four-point structures. The installation of the first cable begins by threading a needle and attaching the double thread to an anchor pin with a knot. I push the needle through the fabric as close to the anchor pin as possible and guide it up towards the top of the mast, working with the elasticity of the fabric.

Outlining the edge catenaries. [8.09]

Here I bring the needle back outside. Two or three winds of the thread around the pin at the top of the mast will hold it in place without totally fixing it. I now push the needle back through the fabric at a spot close to the mast top and in line with the opposite anchor. Again I guide the needle forward under the fabric to the opposite edge, where I pull it out near the anchor pin. I pull on it until the the cable is taut, adjusting the mast, if required. Winding the thread around the anchor pin a few times will usually wedge it sufficiently until it is secured with a few knots. If necessary, a dab of instant glue will make the connection final.

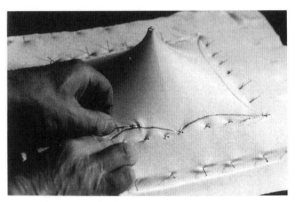

Sewing the edge catenaries. [8.10]

The edge catenaries come next. I start by drawing, with a felt tip pen or a soft pencil, the outline of the edge catenaries on the fabric, all around the structure. It is not critical that the curves are totally even, so a freehand line will be sufficient. The elasticity of the stressed fabric will alone ensure that the catenaries reshape into smooth continuous curves.

For the edge catenaries, I again use a double thread, making small stitches, approximately 3 mm long, and about 2 mm inside the line I have drawn on the fabric. I improve the curvature where I can, not worrying about a stitch getting out of line here or there. After completing one side, I pull the thread tight, so that the catenary curve is just a little shallower than the desired

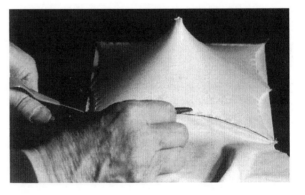

Cutting the edge. [8.11]

Physical and Mathematical Models   203

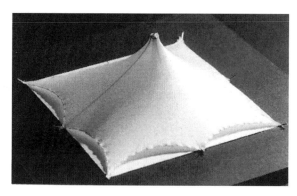

The finished fabric model. [8.12]

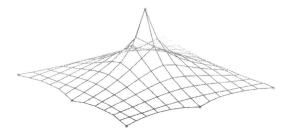

The computer generated model of the same structure. [8.13]

final sag. I then tie it firmly at the anchor pin. This process continues all around the structure, until all edge catenaries are sewn in and anchored.

Now I remove the fabric pins all around the structure, leaving the anchor pins in place. The catenaries will assume slightly larger sags, due to the surface tension which pulls them into tight arcs. The last step is to cut off the loose fabric. I use pointed scissors. They must be sharp and of good quality. Pulling the fabric taut near the spot where I am cutting, I proceed with short cuts along a line just 1 mm outside the thread, careful to avoid cutting the thread, while keeping the cut continuous and smooth. When all excess fabric is removed, all that is left to do is clean up the model.

While stretch fabric models are an excellent means of exploring shapes, testing the behavior of the structure and producing exciting presentation tools, the design of the actual building requires modeling methods of greater precision, which take less effort and are more versatile. Digital models are the answer. Without them, none of the structures illustrated in this book could have been built.

Computer generated digital models serve three particular purposes: generation of the shape of the structure; analysis of its behavior under changing load conditions; and the production of the exact, detailed geometry of each component, especially the cutting patterns of the fabric.

The first and third of these processes, form generation and patterning, are similar, and can be achieved with adaptations of the same mathematical tools. The objective is to find a geometric form for the structure which satisfies the functional requirements of the building while conforming to a prescribed pattern of force flow in perfect equilibrium. The output from this mathematical process must be directly usable to illustrate the structure, to serve as input for the structural analysis, and to dimension the components from which the structure will be built.

Fabric structure designers and fabricators use various form generation programs. Most are based on complex mathematical approaches. I will describe the main features of a process which is simple and direct, and assists in understanding the nature of tensile surface shapes, rather than obscuring it, as mathematical procedures often do. This process takes three distinct steps towards achieving the final shape: the plan shape the isometric shape; and the geodesic shape. I will demonstrate it on the same single-mast tent form used for the stretch fabric model.

Beginning with the plan shape, the first step is to draw a net of force lines which represent the structure. For our purpose, because we want to handle our computations with a common calculator, a square grid of just six elements in each direction is best. To identify each node and each element clearly, we allocate numbers to the rows and columns of the net. If the membrane is to be connected to straight rigid edges, then the plan shape will be complete. If the edges are to consist of catenaries, we shall need to compute their plan shape. I will use this process to introduce the process of iteration, which is the basis for all three levels of this form generation procedure.

Iteration is a mathematical process by which the solution to a problem is found in incremental steps applied to a small portion of the total system. For this approach to succeed, two criteria need to be satisfied: we need to have an accurate mathematical description of the local condition which applies at all times; and every step needs to lead closer to the final solution. Mathematicians call this requirement convergence.

In our case, the applicable equation is derived from the equilibrium of the forces at each node point. Because forces and shapes are related in the most direct way, this derivation is simple and obvious. If we select forces for the net lines and for the horizontal component of the edge catenary, the equation for the sag at each node of the catenary becomes explicit, and the iterative process can now start. We can begin with 0 values at each point, or we can begin with any selection of estimated values. I use an equation which estimates the sag at the center of the catenary, and I then guess the starting numbers for the other points. These starting numbers have no influence on the accuracy of the end result, although they do influence the number of steps in the calculations required to get there. Each step requires application of equation {3}, which consists simply of adding the sags (y) of the two adjacent nodes to the bow factor (Δx_x

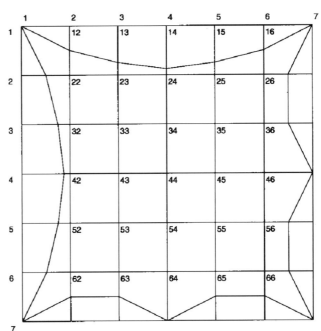

Plan grid of mathematical model. [8.14]

Physical and Mathematical Models

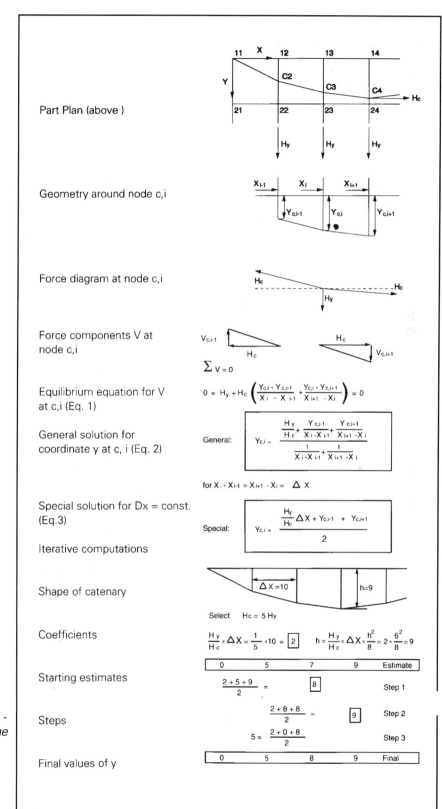

Derivations and computations to determine the shape of the top catenary. [8.15]

Geometric definitions at node i,k	

Force and shape diagrams at node i,k

Equations for force V

$$V_{i+1,k} = H_x \frac{Z_{i+1,k} - Z_{i,k}}{X_{i+1,k} - X_{i,k}}$$

$$V_{i-1,k} = H_x \frac{Z_{i-1,k} - Z_{i,k}}{X_{i,k} - X_{i-1,k}}$$

$$V_{i,k+1} = H_y \frac{Z_{i,k+1} - Z_{i,k}}{Y_{i,k+1} - Y_{i,k}}$$

$$V_{i,k-1} = H_y \frac{Z_{i,k-1} - Z_{i,k}}{Y_{i,k} - Y_{i,k-1}}$$

$$\sum V = V_{i+1,k} + V_{i-1,k} + V_{i,k+1} + V_{i,k-1} = 0$$

General evaluation for z i,k

$$Z_{i,k} = \frac{\left[\dfrac{Z_{i+1,k}}{X_{i+1,k} - X_{i,k}} + \dfrac{Z_{i-1,k}}{X_{i,k} - X_{i-1,k}}\right] Hx_i + \left[\dfrac{Z_{i,k+1}}{Y_{i,k+1} - Y_{i,k}} + \dfrac{Z_{i,k-1}}{Y_{i,k} - Y_{i-1,k}}\right] Hy_k}{\left[\dfrac{1}{X_{i+1,k} - X_{i,k}} + \dfrac{1}{X_{i,k} - X_{i-1,k}}\right] Hx_i + \left[\dfrac{1}{Y_{i,k+1} - Y_{i,k}} + \dfrac{1}{Y_{i,k} - Y_{i-1,k}}\right] Hy_k}$$

Special equation for $z_{i,k}$ for square nets with uniform horizontal forces in both directions.

for $H_x = H_y$ and $S_x = S_y$

$$Z_{i,k} = \frac{Z_{i+1,k} + Z_{i-1,k} + Z_{i,k+1} + Z_{i,k-1}}{4}$$

Derivation for the iterative computation of the isometric shape. [8.16]

Physical and Mathematical Model

Hy/Hc), and dividing this sum by 2. It takes little effort to arrive at the final catenary shape, especially when we realize that we can accelerate the process by making guessed adjustments on the way. The proof, in the end, is that the sag numbers will no longer change, no matter how often we repeat the process. The process is demonstrated on the long catenary. For the short catenaries, only the end results are given.

In the isometric shape we retain the plan geometry and horizontal force components established from the plan shape and allow the node points to move up vertically into a position of three-dimensional equilibrium using a similar iterative procedure. To compute the unknown elevation (z) of anyone node point we have to add the elevation values of the four adjacent node points to the bow factor, and divide by four. If we neglect the weight of the membrane, which is acceptable at least for preliminary shapes, the equation is even simpler: the bow factor becomes zero, and the elevation of the node in question will merely be the average of the four adjacent node elevations. Each step moves the elevation of a node point closer to the final shape, until there is no more movement.

The process is shown on our simple tent form with rigid edges. It is illustrated by a number of isometric images and a table. General equations are also presented, which apply when the net forces in the two directions are not identical, and when the lengths of the net elements vary. The process is not limited to the intersection of two net lines, and this permits the modeling of cables intersecting a node point of a fabric net.

This isometric method of form generation is a quick and simple way of producing and illustrating workable tensile structure surfaces. And it can be a first step to a final design shape and analysis. But inital programs based on this method were not any more user friendly than other formfinding methods.

In the last few years, my son, Ralf Berger, helped convert this iterative method into a user-friendly, design-oriented computer program. He applied his amazing programming capabilities to

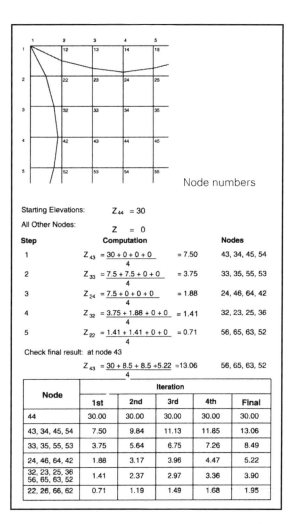

Iterative e computations of z coordinates of the isometric shape. [8.17]

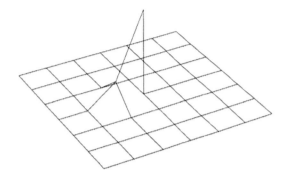

Isometric shape: step 1. [8.18]

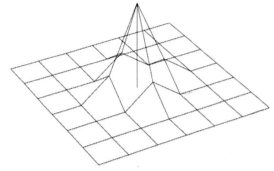

Isometric shape: step 2. [8.19]

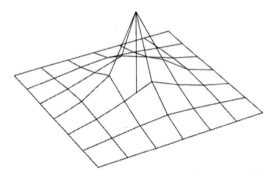

Isometric shape: step 3. [8.20]

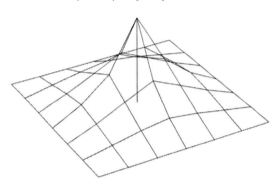

Isometric shape: Final. [8.21]

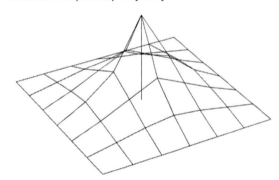

With added ridge cable. [8.22]

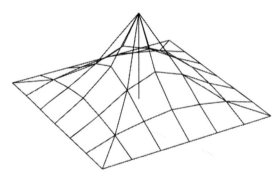

With added corner cable. [8.23]

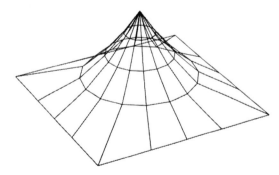

Radial tent shape. [8.24]

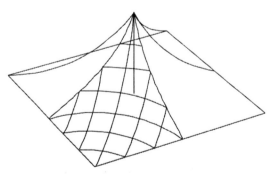

Four point geodesic shape. [8.25]

Physical and Mathematical Models

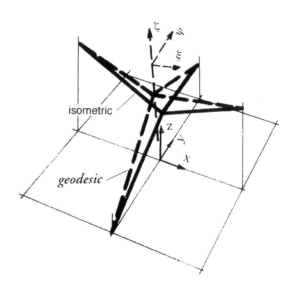

The transition from the isometric to the geodesic shape is achieved by rotating the coordinate system at each node and at each step of the iteration so that the ζ-axis is "normal" (at right angles in each direction) to the surface.
[8.26]

the program, basing the design process on a mesh rather than separate lines. The program, which I call *"SurfaceForm"* is described at the end of this chapter and is illustrated in a few images. It will be available as a Java program on my website, *HorstBerger.com*.

For the final design of large structures the isometric shape does not represent the most practical and economical surface. The sustained stresses (prestress) of surfaces created with the isometric method may vary by as much as +/- 20 %; also, at a node, forces need to be transferred from one line to the intersecting line. If the fabric connects to cables, these forces may need to be considered in the design of their connection. The resulting patterning may not lead to the most economical use of the fabric.

These shortcomings can be overcome by converting the isometric shape into a geodesic shape. This is achieved by extension of the same iterative mathematical method. The first step consists of rotating the coordinate system at a given node in such a way that the axis (now called ζ) of the new coordinate system is at right angles - is "normal"- to the lines which connect the two outer node points in the x direction and the two outer node points in the y direction. The new coordinates ξ and ψ are found at the intersection point of these two lines, in a plane which is parallel to those two lines. Application of the equilibrium conditions derived for the isometric shape to this node in the rotated coordinate system will yield the new coordinate z. Transferring the ξ, ψ, ζ coordinates of this corrected node point back to the x, y, z system completes the sequence for this node. This sequence is now repeated for the next node and, subsequently, for each node in the system, until the coordinates for each and every node point in the net experience no further significant change.

The resulting surface net configuration is geodesic, meaning that the grid lines - strictly speaking, only when their polygon shape is replaced by continuous smooth curves - will intersect at right angles in the plane of the surface at each node. The force along any such geodesic line is constant, permitting the design of fabric membranes with constant prestress levels in the warp and the fill. The resulting surface is a minimum surface, meaning that it has the smallest surface area possible to cover a space with the given fixed support points. More important for the economy of the structure is that the membrane strips of fabric, called "gores" in the fabric industry, resulting from this design approach fit into the parallel boundaries of the long rolls of fabric coming off industrial looms. Geodesic surfaces, therefore, allow patterning of the membrane fabric with a minimum of loss of material.

Economic patterning requires that the net lines are properly spaced along the edges, and along cables where

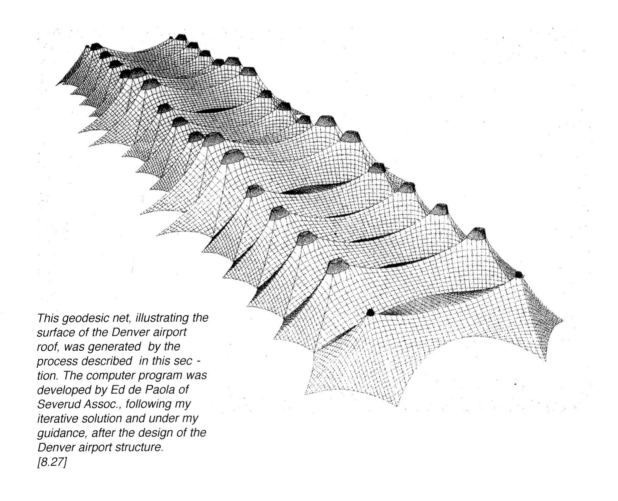

This geodesic net, illustrating the surface of the Denver airport roof, was generated by the process described in this section. The computer program was developed by Ed de Paola of Severud Assoc., following my iterative solution and under my guidance, after the design of the Denver airport structure.
[8.27]

Physical and Mathematical Models

fabric panels are spliced. Good form generation programs require built in controls which allow the predetermination of net spacing along such lines. For fabric structures, geodesic shaping with controlled edge spacing leads to the best structural forms.

The stress analysis of tensile structures requires totally different mathematical methods. The equilibrium shape, as produced by the various methods discussed, is simply a starting point. Under load, tensile surface structures experience large deflections. Their change of shape is a major response to the load, helping resist it in combination with a change of stress. The behavior of these flexible structures, and their capacity to carry load, cannot therefore be analyzed using linear programs based on totally elastic behavior and the insignificant form changes of rigid structural systems.

Nonlinear analysis programs capable of predicting stresses and form changes correctly are many times more complex than linear, elastic pro-grams. We are not yet capable of describing non-linear events - which belong to the new physics of "chaos" - using simple algorisms, and we there-fore need to adapt linear methods in programs which apply loads and shape changes in iterative steps, finding the new form gradually. In the process, some net lines may go slack, reverse their curvature, and then pick up stress again. Non-linear analysis programs need to be able to simulate this complex behavior accurately.

The analysis program used for the structures described in this book was developed by Dr. William Spillers, then of Columbia University and now Chairman of the Department of Civil Engineering at the New Jersey Institute of Technology, a pioneer in nonlinear analysis of structures. It has gone through many steps of in-house improvements, mostly by former students of Dr. Spillers. Its results, confirmed by detailed measurements on full scale prototype structures, have been found to predict stresses and deformations very closely.

The history of this program's use is a fascinating measure of the evolution of computer capacity. The analysis of the Jeddah structure, in 1979, required a Cray computer, then the largest, most powerful and most expensive computing machine available. Today, a $2,000. PC or Macintosh can handle this task with ease.

In the first edition of this book I wrote: "There is

also doubt in my mind that the turbulent progress in digital computing and drafting devices currently revolutionizing architectural and engineering offices will drastically change the attitude of design professionals towards the use of geometrically complex forms of surface structures. The new digital tools will empower the designers of the built environment to include surface structures in their design vocabulary, so that such structures will soon become familiar images in the visual landscape."

The revolution I was talking about is in full progress. Among students of architecture the shift to CAD programs as the major design tool has so far produced more confusion than understanding. It empowers them to produce enticing images often not backed by reality of space, especially buildable space. Two-dimensional images of three-dimensional space are easily misleading. For buildings whose structure depends on form for its efficiency, digital tools are needed which can help the designer to make make meaningful choices, even without the extensive engineering training which would be needed to have the required judgement.

My "SurfaceForm" program, described here, has this objective in mind. It is simple enough to be used by any architectural designer, and it is powerful enough to be a powerful tool even for

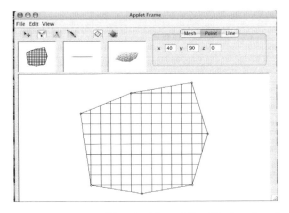

The desktop of the "Surfaceform" program with the plan of a structure outlined. [8.28]

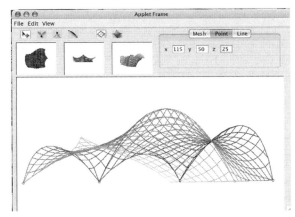

By adding arches and catenaries, the structure takes on its shape. [8.29]

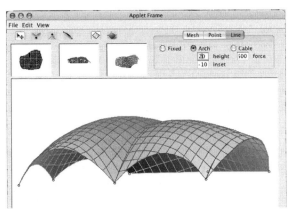

The structure can be converted into a grid shell by adding a curvature factor and adjusting some of the arches. [8.31]

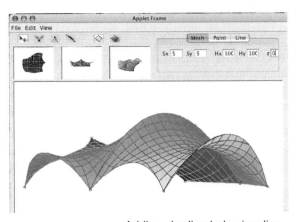

Adding shading help visualize the form. [8.30]

Physical and Mathematical Models

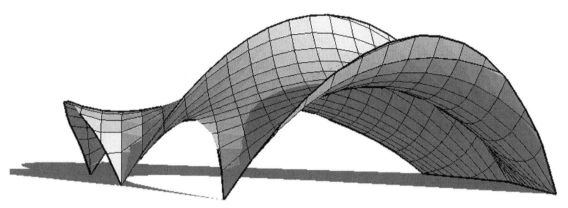

The end result of the Surfaceform design process can be saved as a dxf document and imported into CAD projects. The image above is a perspective view of the structure shown before. It is shown as a "Sketch-up" project.
[8.32]

experienced and advanced designers, like me.

The work window of the of "SurfaceForm" lets you look at the project in plan, section, and isometric view. There are a number of tools to be selected in the menu bar.

It makes sense to start with the plan. The window shows you a mesh of ten squares in each direction, with an x and y-spacing of 10 (feet, meters, whatever you like, just stay consistent). You can shape the outline of this mesh by adding and moving node points along the edges. These points can have various elevations, and they can be origins of support lines, which are straight lines, arches (both vertical or sloping inward or outward), or catenaries.

The plan can be refined by adding interior support points or support lines, and by changing the mesh in each of the two directions. Interior lines can be arches or cables. Arches are parabolic between support points. If different lines or curvatures are required, intermediate points can be interjected as spring points for arches with varying curvatures. The support points of these members are visible in form of small circles, reminding the designer, that something has be done by a condition which is not automatically handled by the funicular shape of the structural system. (The circles can be eliminated for presentation purposes).

The shape can be be saved in various ways, including as "dxf" document, allowing it to be introduced into the design of other applications, including most CAD programs. Here they can be shown as perspectives, transformed into actual building designs, or added as components to buildings under design.

More information is available on my website, *HorstBerger.com.*

214 Light Structures
Structures of Light

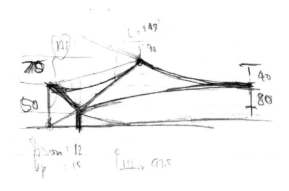
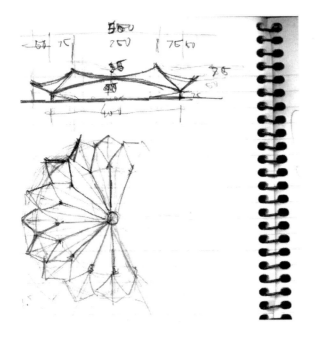

These sketches for an arena are from a 1987 notebook. I drew them on an airplane from London to New York.
Some of my best ideas came on airplanes. This is a good concept and still valid. Tents are suspended from a ring of cantilevering A-frames which are tying back to a second ring of A-frames and down to the foundations. As the Riyadh stadium demonstrated, this is a highly economical and dramatic way of creating a beautiful huge space.
[9.01]

9
Ideas and Possibilities

An article in the May 1980 issue of Architectural Record, was entitled "Tent Structures: Are They Architecture?" and featured the Haj Terminal structure in Jeddah, Saudi Arabia, and the Florida Festival Building, at Sea World, in Orlando, Florida. Its author, the late Robert E. Fisher, senior editor of the magazine, was in no doubt about the answer; his many articles on tensile structures, beginning in 1975, were instrumental in making the architectural profession aware of the power and beauty of this new architectural technology. When my dear friend, Bob Fisher, died of cancer in 1984, the Jeddah project had been completed, the Riyadh Stadium was under construction, and the San Diego Convention Center was on the drawing boards. Bob would have seen the Denver Airport structure as confirmation of his faith in the viability of this new way of making practical, novel, and beautiful architecture.

While the structures completed in the three last decades of the 20th century demonstrate the success of tents as architecture, it is the many unbuilt designs and the untried possibilities which suggest a much broader scope for fabric structures in the built environment. Not only can we envision richer and gentler architectural spaces and structures with combination of enticing new forms, but the characteristics of fabric (translucency,

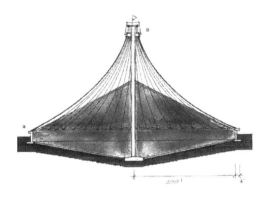

This 400 ft. dia. tent has the ideal shape for enclosing grain storage, as this design shows. A flexible structure is safer against explosion.
[9.02]

reflectivity) can help create a built environment much friendlier to nature. Beyond that, as lower cost fabrics and simpler construction technologies appear, new economies may make it possible to share the benefits of creative architectural design with a much larger part of the world's population. The following examples are limited to examples of my own experience.

There are a number of industrial applications for which tensile structures offer clear advantages. Enclosures for grain storage facilities are the most obvious. The disastrous explosion of a concrete grain silo had motivated me to investigate a fabric structure solution. I found a tent-shaped tensile structure to be the logical form for enclosing the cone shape that grain or other granular materials naturally assume. In my proposed design, the central support column is a cylinder which also serves as the grain elevator. The most damage an explosion can cause is a rip in the fabric. Due to the flexible nature of the structure, however, even this is unlikely. Reflectivity keeps the space from overheating in the sun; translucency helps those working inside. Inert fabrics, such as Teflon or silicon coated materials, can be used to cover chemically aggressive materials. The much more economical polyethylene fabrics may be acceptable and may make very large storage facilities possible with simple, attractive structures.

Coal fired power stations require huge coal piles, where rain water, seeping through the coal into the soil below, will cause ground water pollution. Studies carried out by Horst Berger Partners for the Electric Power Research Institute (EPRI) demonstrated the effectiveness and economy of fabric structures as covers for pollution prevention. All or part of the cost of building such covers is compensated by increased energy efficiency resulting from improved moisture control of the coal. A design proposed for such a structure to cover a New York State Electric and Gas Company (NYSEG) plant near Binghamton, N.Y., is shaped like a large umbrella. It shows that the visual appeal of tensile structures could help transform industrial facilities of this kind from eyesores into attractive additions to the landscape.

Cooling towers, long an interest of mine, were also part of the studies for EPRI. Their purpose is to increase the efficiency of the power generating plant and reduce thermal pollution of lakes or rivers. Thin-shell concrete structures have served this purpose for many decades.

This umbrella shaped roof was designed to cover a coal storage pile for an electric power generating plant.
[9.03]

Cable-restrained fabric cooling tower design would enable the use of "dry" cooling systems, further reduce pollution and increase the power plant's efficiency.
[9.04]

Ideas and Possibilities

Rendering of cooling tower design for a geothermal power station in Utah. This was part of a study for NREL in 2001. The tower is 200ft. height.
[9.05]

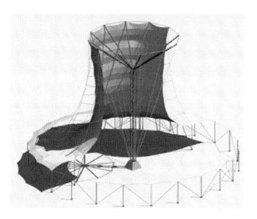

Schematic rendering show the support system of the Utah cooling tower design.
[9.06]

Their hyperboloid shape allows towers to reach 500 ft. height, with a wall thickness of only 8 in. This means that, in proportion to their overall size, concrete cooling tower walls are thinner than egg shells. Nevertheless, our studies demonstrated that fabric construction could be more efficient, especially for "dry" cooling towers, which work on the same heat exchange principles as car radiators, moving cool air past an intricate arrangement of fins, through which a special cooling fluid circulates. Dry towers could further increase the efficiency of the generating plant and eliminate the moisture pollution caused by "wet" cooling towers, in which cooling water drips into the air flow, causing a large amount of the moisture to be carried up by the air and dumped outside on the surrounding landscape. By separating the structural shape of the support cables from the hydraulic shape of the funnel, called a "vail", both air-flow and structural efficiency can be optimized.

In a study for NREL (National Renewable Energy Laboratory) I designed a 200 ft. high fabric cooling tower for a geothermal power plant in Utah. Using the most recently developed polyethylene fabric, this tower would prove to be highly economical. Its design should make it an acceptable structure in the landscape of Utah.

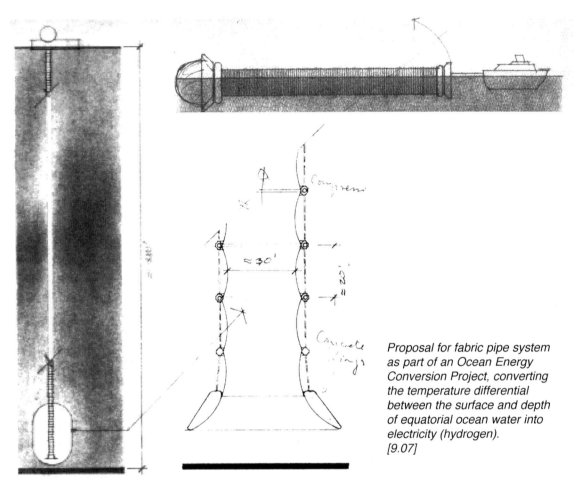

Proposal for fabric pipe system as part of an Ocean Energy Conversion Project, converting the temperature differential between the surface and depth of equatorial ocean water into electricity (hydrogen).
[9.07]

A fabric structure would also be the only practical solution for the 3000 ft. long tube proposed in a scheme for converting the temperature difference between the surface and depths of equatorial oceans into electric energy. In my design, concrete rings preserve the circular shape of the fabric funnel against the enormous water pressure differential. For transportation, the tube structure is contracted so that the rings touch and form a solid concrete tube. This concrete barge is then tugged into position, where it is upended, and the tube deployed.

Other unbuilt designs demonstrate the usefulness of fabric for applications which deal with the control of the living environment: a retractable roof system to cover 102nd Street in Edmonton, Canada, was intended to transform this downtown street into an attractive urban space in the long northern winters, without sacrificing the joy of open air and sunshine in the two or three months of the short summer. More ambitious were proposals to cover an entire city farther north, where work-

Ideas and Possibilities 219

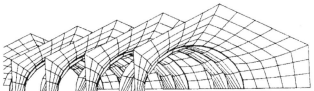

The model photograph [9.08] illustrates the retractable roof cover for 102nd Street in Edmonton, the northernmost city of Canada. The computer perspective shows the shape of this roof membrane. [9.09]

ers extract oil from shale in outdoor air temperatures of 500 F. Here a somewhat larger scale Jeddah airport module concept would allow to built a few modules first, then use them to, step by step, "grow" to the city.

The design for the winter enclosure of the ground level of 17 State Street, at the tip of Manhattan Island in New York, is another variation of this concept. This handsome 45 story office building sits on 30 ft. high columns, and generously offers to the public all of the street level except for a small, circular glass-enclosed lobby. The idea was to protect the ground level against the icy winds which blow across the bay into the narrow streets of downtown Manhattan. A series of overlapping fabric "leaves" attach to light aluminum tube frames, small units that are easy to ship, install, remove, and store. With their organic shapes, I wanted to give a soft edge to the plain lines of this building.

Fabric structures make good transitions between the exterior environment and the interior space of a building. My design of a large canopy for the rear of the Governor's Mansion in Frankfort, Kentucky, is a simple example. It links the formal house to the attractive gardens, offering a shaded space that is large enough to shelter the annual garden party after the Kentucky Derby.

A composition of fabric roof elements for a low-rise hotel proposal in Arizona takes this concept a step further: a sweep of sidewalk canopies in front of the hotel

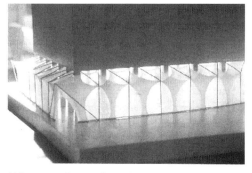

Winter enclosure for 17 State Street in New York City. Leaf-shaped fabric panels are easily installed or removed. [9.10]

links the various conference rooms to the entrance canopy and the fabric covered lobby. Behind the lobby, the fabric roof extends to cover a pool and the walkways to the hotel rooms.

My design of the entrance canopy for the offices of Birdair Inc. is more modest - but it was built. For Birdair, the world leader in fabric construction, a fabric canopy was both practical and symbolic.

Garden canopy for the Governor's mansion in Frankfort, Kentucky. [9.11], above, and fabric enclosure for a hotel design in Arizona [9.12] show transition structures, linking the building to the space around it.

For the shaping of interior spaces, fabric membrane shapes offer amazing possibilities. An example is the study model for one of the many by-centennial projects, this one in Washington, D.C.

The only fabric project I have in the City of New York is the enclosure for the Whale Pool at the Aquarium. (New York is the city which has been my operating base for 35 years and in which I can point to numerous conventional projects for which I was the structural engineer, some of them of substantial scope).

Sliding glass walls make the whale pool enclosure convertible. In the summer, the roof provides shade for the light sensitive whales; in the winter, it is an enclosed and heated space for whales and visitors alike.

The entrance canopy at the office of Birdair Inc. not only guides the visitor into the building, but is also symbolic of the companies activity and commitment. [9.13]

Architecture is space. Fabric structures shape space in ways never before possible. Model for "We the People" pavilion for the by-centennial cele- bration in Washington, DC. [9.14]

Ideas and Possibilities 221

The roof cover for the New York Aquarium's Whale Pool is the first and - so far - the only of my fabric structure designs in the city. The prophet is rarely heard in his own home town. 1994.
[9.15]

The enclosure structures for the New York Aquarium's amphitheater were designed, approved by the City, but cancelled by the mayor. 2001.
[9.16]

The support system consists of arches with constant radius, which - therefore - become flatter as their span gets smaller in the trapezoidal roof plan. The combination of a translucent fabric roof and vertical glass surfaces, makes for an especially friendly space.

Unfortunately, I was not involved with the construction phase of the project. The Silicone coated fiberglass fabric selected, proved to be worse with regard to collecting dirt on the fabric. Though the translucency inside is little reduced, the outside appearance suffers.

The hope to replace the fabric and complete aspects of this building was connected to the next project at the Aquarium: the roofing of the amphitheater. The design of this project was completed, approved by the City, and out for bidding, when it was cancelled by the mayor in the wake of September 11. This is an unfortunate setback for the Aquarium. Keeping the sun off the 1500 aluminum seats in the summer and enclosing the amphitheater on the winter would have been a big boost for the Aquarium.

These examples of largely unexplored design possibilities suggest that fabric structures can greatly enrich our built environment with forms that are highly adaptive in space and time. Equally exciting is the potential for shaping building forms and spaces in ways which are organic and sculptural, shedding the straightjacket of the rectangular grid, symbolic of a regimented, imprisoned

society. The model study for a bicentennial pavilion in Washington, D.C., illustrates the potential for shaping interior spaces. The design for a wall to hide a proposed coal pile in Hoboken, N.J. indicates that the use of fabric structures for vertical surfaces, including building facades, is a vast field of application still open to exploration. A number of models by my students, and forms

Architecture is also sculpture. This fabric wall design had the purpose of screening a huge coal pile planned for Hoboken, N.J. Its shown in a perspective sketch [9.18] above and a partial stretch fabric model [9.19] at left.

made with my new form-finding program show some formal possibilities.

There are still obstacles to the wider use of fabric structures; and the complexity and novelty of this new technology are among them. But with the rapidly increasing use of the computer, the design of fabric structures will become easy. My new form-finding program, described in Chapter 8, suggests such a develop-

One of many study models made by my students at the School of Architecture of CCNY, part of the City University of New York, where I have been teaching for the last 15 years. [9.20]

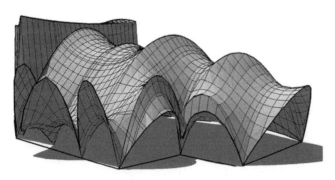

My schematic theater design for the Woodlands in Texas, using my new form-finding method, transferred by dxf to Sketch-up. These are forms easily built, using fabric membrane surfaces. [9.21]

Ideas and Possibilities 223

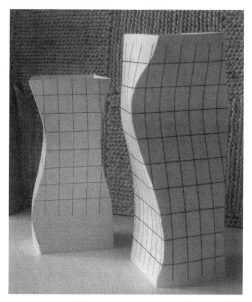

High-rise wall study, showing building surfaces which could be build in fabric or even from rectangular rigid panels, since the entire volume is made from a rectangular sheet. The outermost layer of my own body, and yours, is skin, after all.
[9.22]

ment. With new fabric materials becoming available and as construction becomes more widespread, costs will drop substantially. The visibility of public projects such as the Denver Airport, the Dan Diego Convention Center, and Canada Place, will make fabric architecture acceptable and desirable to a growing number of people. Ultimately, however, I believe that such acceptance will depend on society's readiness to relinquish the rigidities inherited from a past whose static formulas are no longer adequate in a dynamic new world. We are witnessing the failure of large hierarchies in government, industry, and commerce because of their inflexibility. In the built environment, vacancies are increasing in high-rise towers built to last a thousand years, while the frozen street grids they occupy have turned into canyons of anonymous crime.

Our survival depends on recreating communities whose members feel responsible for their actions and for each other, respect the natural environment we have been born into, and understand the unity of all life. It appears that ancient communities were based on such respect and understanding. The soft forms of their buildings were expressions of this spirit. Fabric structures, with their gentle forms obeying a higher order, offer a small step in this daring process of moving forward into the past.

My most recent design: a very large span enclosure for a auto-racetrack. The sketch shows three interior bays only.
[9.23]

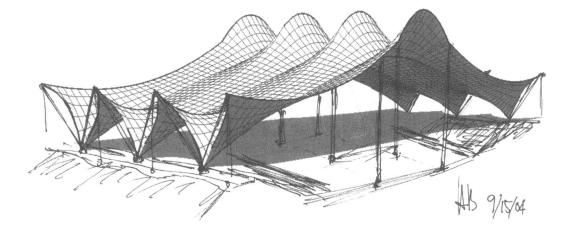

**Light Structures
Structures of Light**

Project Credits

225

Project Name Completion	Project Architect, Principal Designer	Roof Designer and Structural Engineer	Roof Fabricator- Contractor	Additional Information
Cohen Stadium Roof El Paso, Texas, USA, 2003	Ron Brown Architects	Horst Berger Partners DeNardis Engineers	Birdair	Initially designed by Horst Berger Partnes in 1989
Amphitheater Roof and Pool Enclosure New York Aquarium, New York, N.Y. Approved and bid 2003	Goldstone and Hinz, Architects New York, N.Y.	Horst Berger with DeNardis Associate		This project was shelved by the Mayor after 9/11
Unidome Roof Replacement University of Northern Iowa Cedar Falls, Iowa, USA, completted 1999	Light Structures Design Consultants with Horst Berger White Plains, New York, USA	Light Structures / Horst Berger	Koit-Hightech	
Enclosure of Whale Pool New York Aquarium, New York, USA, Completion Summer 1996	Goldstone and Hinz, Architects New York, New York, USA	Severuds Associates, P.c. New York, New York, USA.	DCI, Inc. Norcross, Georgia, USA	
Jepesson Terminal, Denver International Airport Denver, Colorado, USA, 1994	c.w. Fentress J.H. Bradburn & Associates Denver, Colorado, USA	Severud Associates, Horst Berger, Principal Consultant	Birdair Inc. Amherst, New York, USA	
Birdair Entrance Canopy Amherst, New York. USA, 1991	Horst Berger	Horst Berger with the Engineers of Birdair Inc.	Birdair Inc.	
San Diego Convention Center San Diego, California, USA, 1990	Arthur Erickson Assoc., Vancouver, B.C., Canada	Horst Berger Partners	Birdair Inc.	
The Cynthia Woods Mitchell Center for the Performing Arts The Woodlands, Texas, USA, 1990	Horst Berger Partners in Association with Sustaita Associates, Houston, Texas, USA	Horst Berger Partners	Birdair Inc.	Consultants during conceptual design: Jaffe Acoustics, John Toffoli, Ellerbe-Becket
Wimbledon Practice Facility Wimbledon, UK, 1988	Ian C. King, Architect London, UK	Horst Berger Partners	Koit Herbert Koch GmbH Rimsting, Germany	Structural Consultants: De Leuw Chatwick, London, UK
McClain Football Practice Facility The University of Wisconsin Madison, Wisconsin, USA, 1987	Bowen, Williamson, Zimmermann, Architects Madison, Wisconsin, USA	Horst Berger Partners	Birdair Inc.	
Shoreline Amphitheater Mountain View, California, USA, 1986	The Bluerock Partnership Newport Beach, California, USA	Horst Berger Partners	ODC Inc. Norcross, Georgia, USA	Teflon roof replacement by Birdair Inc.
The Great Hall of Alexandra Palace London, UK, 1986	The Alexandra Palace Development Team Peter Smith, Director	Horst Berger Partners	Silicon Fabric by ODC, Fabrication and construction by Koit	
Winter Enclosure Structures at 17 State Street New York, New York, USA Design 1986	Horst Berger Partners	Horst Berger Partners		Architects for 17 State Street: Emery Roth & Sons New York, New York, USA
King Pahd Stadium Riyadh, Saudi Arabia, 1985	Ian Fraser, John Roberts & Partners, London, UK	Geiger Berger Assoc. P.c. New York, New York, USA	Birdair Inc.	Construction phase engineering: Schlaich & Partner, Stuttgart, Germany

226 Project Credits Continued

Project Name Completion	Project Architect, Principal Designer	Roof Designer and Structural Engineer	Roof Fabricator-Contractor	Additional Information
Canada Place Vancouver, B.C. Canada, 1985	Z-dler Roberts Partnership, Toronto, Canada	Geiger Berger Assoc.	Birdair Inc.	Structural engineers for substructure: Read, Jones, Christoffierson Vancouver, B.C., Canada
Crown Center Pavilion Kansas City, Missouri, USA, 1984	Geiger Berger Assoc.	Geiger Berger Assoc.	ODC	
Sibley Horticultural Center Callaway Gardens, Pine Mountain, Georgia, USA, 1984	Craig, Gaulden & Davis, Architects, Inc., Greenville, S. Carolina, USA	Geiger Berger Assoc.	Owens-Corning Fiberglas Corp., Toledo, Ohio with ODC	Principal Designer of the Center: Robert E. Marvin, PA, Landscape Architect, Walterboro, S. Carolina, USA
Tennessee Amphitheater Knoxville, Tennessee, USA, 1982	McCarty, Bullock & Holsaple Inc., Knoxville, Tennessee, USA	Geiger Berger Assoc.	Owens-Corning Fiberglas Corp. Toledo, Ohio, USA	
Ashboro Zoo Ashboro, N. Carolina, USA, 1982	Hayes & Howell Southern Pines, N. Carolina, USA	Geiger Berger Assoc.	Owens-Corning Fiberglas	
The Mall at 163rd Street North Miami, Florida, USA, 1982	Gale, Kober Associates Los Angeles, California, USA	Geiger Berger Assoc.	Owens-Corning Fiberglas	
Haj Terminal of the Jeddah International Airport Jeddah, Saudi Arabia, 1981	Skidmore Owens & Merril Chicago, Illinois, USA	Skidmore, Owens & Merril, Geiger Berger Assoc.	Owens-Corning with Birdair Inc.	Geiger Berger was consultant to SOM in the conceptual stage and the engineer of the final design of the fabric structure
Bullock's Fashion Island Department Store, San Mateo, California, USA, 1981	L. Gene Zellmer Associates Fresno, California, USA	Geiger Berger Assoc.	Birdair Inc.	Interior Architects: Environmental Planning and Research Inc. San Francisco, California, USA
Picnic Pavilion at Seaworld in San Diego, California, USA, 1982	Geiger Berger Assoc.	Geiger Berger Assoc.	Birdair Inc.	
Florida Festival Seaworld of Orlando, Florida, USA, 1980	Robert Lamb Hart Assoc., New York, New York, USA	Geiger Berger Assoc.	Birdair Inc.	
NBC-News Center for the 1980 Democratic Convention New York, New York, USA, 1980	Geiger Berger Assoc.	Geiger Berger Assoc.	Airtech Industries Inc. Ruthedord, New Jersey, USA	
Bullock's Department Store San Jose, California, USA, 1979	Environmental Planning and Research Inc. San Francisco, California, USA	Geiger Berger Assoc.	Owens-Corning with Birdair Inc.	
Queeny Park Pavilion St. Louis County, Missouri, USA, 1978	Jones Mayer Associates St. Louis, Missouri, USA	Geiger Berger Assoc.	Birdair Inc.	
Bandstand Structure Jacksonville, Florida, USA, 1978	William Morgan Architects Jacksonville, Florida, USA	Geiger Berger Assoc.	Birdair Inc.	Concept design of building by the Warner LeRoy Studio
Bicentennial Pavilions, Philadelphia, Pennsylvania, USA, 1976	H2L2, Architects/Planners Philadelphia, Pennsylvania, USA	Geiger Berger.Assoc.	Owens-Corning Fiberglas with Birdair Inc.	
Store Structures at Great Adventure New Jersey, USA, 1973	Geiger Berger Assoc.	Geiger Berger Assoc.	Birdair Inc.	

Illustration Credits

The photographs not identified below were taken by the author or are part of his
collection. Hand drawings not identified were drawn by the author. CAD drawings were
produced by Light Structures Design Consultants, by Mimi Kueh or by the author.

| 1 | **Introduction - Building a Gentler Environment** |

1.02 Hans Joachim Seidel
1.11 Martin Brown

| 2 | **In the Beginning: Domes** |

2.01 Labelle Prussin
2.02 Henry de Lumley, Directeur du Museum National d'Histoire Naturelle, Paris
2.03 From Native American Architecture (NAA) by Peter Nabokov and Robert Easton. Copyright 1988 by Peter Nabokov and Robert Easton. Reprinted by permission of Oxford University Press, Inc. Drawing by Robert Easton
2.04 NAA, Robert Easton
2.05 Smithsonian Institutions (J.Mooney)
2.08 Slide Library of the School of Architecture, City College of New York (CCNY)
2.11 Labelle Prussin
2.12 Collection Musee de L'Homme, Paris
2.14 CCNY
2.16 CCNY
2.18 Bethlehem Steel Corporation
2.19 Felix Candela
2.20 CCNY
2.21 Heinz Isler

| 3 | **From Tents to Tensile Architecture** |

3.01 Moroccan Tourist Office
3.02 Labelle Prussin
3.03 The Tent Book (TTB) by E.M. Hatton, Published by Houghton Mifflin
3.04 TTB
3.05 TTB
3.06 W.H.Over State Museum, Vermillion, South Dakota
3.07 Montana Historical Society
3.08 Staatliches Museum zu Berlin, Preußischer Kulturbesitz, Museum fur Völkerkunde, Foto: Dietrich Graf
3.09- NAA, Robert Easton
 3.10
3.11 CCNY
3.13 Labelle Prussin
3.14 TTB
3.15 Rainer Graefe
3.16 National Maritime Museum, Greenwich, London
3.19- Science & Civilization in China (SCC) Vol. IV:3
 3.20 by Joseph Needham, Reprinted with the permission of Cambridge University Press
3.21 CCNY
3.23 CCNY
3.28- CCNY
 3.30
3.31- Bethlehem Steel Corp.
 3.32
3.35 CCNY
3.37- Mamoru Kawaguchi
 3.38 CCNY
3.40- Frei Otto
 3.42

| 4 | **How Tensile Structures Work** |

4.02 Bethlehem Steel Corp.
4.19 Birdair

| 5 | **Materials for Tensile Structures** |

5.02 Attila Rona
5.03 Birdair
5.05 Birdair
5.07 Robert Reck / Birdair
5.17 Owens/Corning
5.18- Birdair
 5.20
5.24 NovaShield
5.25 Shade Sails

6　20 Years of Fabric Structures: 1973-1993

6.02　Wayss & Freytag
6.06　Birdair
6.1.05　Brian Green/National Geographic Society
6.2.16　Ann Yaw
6.3.02　Owens Corning
6.3.03　Skidmore, Owens & Merrill
6.3.06　DeNardis
6.3.10　Robert Azzi / Woodfin Camp
6.3.27- Jay Langlois / Birdair
6.3.30
6.4.02　Robert Fisher
6.5.05　Steven Proehl
6.5.16　Russell Abraham
6.6.13- Birdair
6.6.15
6.6.17　Birdair
6.7.05　Owens Corning
6.7.09　Eric Oxendorf
6.7.13　Ian King
6.8.11　Robert Reck / Birdair
6.9.16　Robert Reck / Birdair
6.10.01　Robert Reck / Birdair
6.10.03　Fentress/Bradburn
6.10.08　Birdair

7　Covering Very Large Spaces

7.08　ENR
7.15　Birdair
7.20　DeNardis

9　Ideas and Possibilities

9.05　DeNardis
9.06　DeNardis
9.16　DeNardis

Made in the USA
San Bernardino, CA
10 July 2014